Nellie Blanchan (De Graff) Doubleday

Birds that Hunt and are Hunted

Life Histories of one Hundred and Seventy Birds of Prey, Game Bids and

Water-Fowls

Nellie Blanchan (De Graff) Doubleday

Birds that Hunt and are Hunted
Life Histories of one Hundred and Seventy Birds of Prey, Game Bids and Water-Fowls

ISBN/EAN: 9783744750714

Printed in Europe, USA, Canada, Australia, Japan

Cover: Foto ©berggeist007 / pixelio.de

More available books at **www.hansebooks.com**

BIRDS THAT HUNT AND ARE HUNTED
LIFE HISTORIES OF ONE HUNDRED AND SEVENTY BIRDS OF PREY, GAME BIRDS AND WATERFOWLS

BY

NELTJE BLANCHAN

AUTHOR OF "BIRD NEIGHBORS"

WITH INTRODUCTION BY

G. O. SHIELDS (COQUINA)

AND FORTY-EIGHT COLORED PLATES

NEW YORK
DOUBLEDAY & McCLURE CO.
1899

PASSENGER PIGEON.
½ Life-size.

TABLE OF CONTENTS

	PAGE
Introduction by G. O. Shields	vii
Preface	ix
List of Colored Plates	xi
Part I. Water Birds	1
Diving Birds	3
The Grebes	8
The Loons	14
Auks, Murres, Puffins, etc.	18
Long-winged Swimmers	27
Jaegers and Skuas	32
Gulls	35
Terns, or Sea Swallows	46
Skimmers	59
Tube-nosed Swimmers	63
Shearwaters	67
Petrels	68
Fully Webbed Swimmers	73
Cormorants	77
Plate-billed Swimmers	81
Mergansers, or Fishing Ducks	87
River and Pond Ducks	93
Sea and Bay Ducks	114
Geese	134
Swans	143

Table of Contents

	PAGE
PART II. WADING BIRDS	147
Herons and their Allies	149
Ibises	153
Wood Ibises and Storks	155
Herons and Bitterns	157
Marsh Birds	169
Cranes	174
Rails	177
Gallinules	184
Coots	186
Shore Birds	189
Phalaropes	196
Avocets and Stilts	198
Snipe, Sandpipers, etc.	201
Plovers	237
Surf Birds and Turnstones	249
Oyster-Catchers	251
PART III. GALLINACEOUS GAME BIRDS	255
Bob Whites, Grouse, etc.	261
Turkeys	288
Columbine Birds	291
Pigeons and Doves	294
PART IV. BIRDS OF PREY	299
Vultures	304
Kites, Hawks, Eagles, etc.	309
Barn Owls	335
Horned and Hoot Owls	337
INDEX	353

INTRODUCTION

BIRD life is disappearing from the United States and Canada at so alarming a rate I sometimes feel it is wrong, at this day and age of the world, to encourage the hunting and shooting of birds of any kind. Mr. W. T. Hornaday, the Director of the New York Zoölogical Society, has recently collected and compiled statistics from more that thirty states, showing that the decrease of birds within the past fifteen years has averaged over forty per cent. At this rate another twenty years would witness the total extermination of many birds in this country. Several species have already become extinct, and others are rapidly approaching the danger line. Conspicuous among these are the wild turkey and the pinnated grouse, two of the noblest birds on the continent. Several species of water-fowl are also growing scarce.

Not only are game birds pursued and killed, in season and out of season, under the name of sport and for market, but the song birds, plumage birds, water-fowl, and many innocent birds of prey are hunted, from the Everglades to the Arctic Circle, for the barbaric purpose of decorating women's hats. The extent of this traffic is simply appalling. Some of the plumes of tropical and semi-tropical birds sell at as high a price as fifteen dollars an ounce. No wonder the cupidity of ignorant and heartless market hunters is tempted by such prices to pursue and kill the last one of these birds. It seems incredible that any woman in this enlightened and refined age, when sentiment against cruelty to animals is strong in human nature, could be induced to wear an ornament that has cost the life of so beautiful a creature as an egret, a scarlet tanager, or a Baltimore oriole. What beauty can there be in so clumsy a head decoration as an owl or a gull? Yet we see women whose nature would revolt at the thought or the sight of cruelty to a horse or a dog, wearing the wings, plumes, and heads, if not the entire carcasses of these birds. Not only is the life of the bird sacrificed, whose plumage is to be thus worn, but in thousands of instances the victim is the mother bird, and a brood of young is left to starve to death in consequence of her cruel taking off. Is it not time to check this ruthless destruction of bird life by the enactment and enforcement of proper laws?

Introduction

A great crusade against bird slaughter is sweeping over the country. Thousands of progressive educators have inaugurated courses of nature study in the schools, which include object lessons in bird life. Bird protective associations are being formed everywhere. The League of American Sportsmen is doing a noble work in this direction. It is waging a relentless war on men who kill game birds out of the legal season, or song birds at any time. This organization stands for the highest type of men who hunt, and it is laboring to educate the other kind up to its standard. The surest way to promote this sentiment of bird protection is to induce our people to study the birds. Nearly every man, woman, and child who becomes intimately acquainted with them learns to love and to respect them for their incalculable benefits to mankind. The reading of such a book as this is a step in the right direction. The next step should lead the reader into the fields, the woods, and by the waters.

I have read the manuscript of this book carefully. It shows the most patient and industrious research, and it is safe to say no work of its class has been issued in modern times that contains so much valuable information, presented with such felicity and charm. The author avoids technicalities, and writes for the layman as well as for the naturalist. While the volume caters in a great measure to sportsmen, yet it is the hope of the author and the editor that they may learn to hunt more and more each year without guns; for all true sportsmen are lovers of nature. The time has come when the camera may and should, to a great extent, take the place of the gun. Several enthusiasts have demonstrated that beautiful pictures of wild birds may be made without taking their lives. How much more delight must a true sportsman feel in the possession of a photograph of a beautiful bird which still lives than in the mounted skin of one he has killed! A few trophies of this latter class are all right, and may be reasonably and properly sought by anyone; but the time has passed when the man can be commended who persists in killing every bird he can find, either for sport, for meat, or for the sake of preserving the skins.

The colored plates in this book are true to nature, and must prove of great educational value. By their aid alone any bird illustrated may be readily identified.

G. O. SHIELDS.

PREFACE

THE point of view from which this book and "Bird Neighbors" were written is that of a bird-lover who believes that personal, friendly acquaintance with the live birds, as distinguished from the technical study of the anatomy of dead ones, must be general before the people will care enough about them to reinforce the law with unstrained mercy. To really know the birds in their home life, how marvelously clever they are, and how positively dependent agriculture is upon their ministrations, cannot but increase our respect for them to such a point that wilful injury becomes impossible.

In Audubon's day flocks of wild pigeons, so dense that they darkened the sky, were a common sight; whereas now, for the lack of proper legislation in former years, and quite as much because good laws now existing are not enforced, this exquisite bird is almost extinct, like the great auk which was also seen by Audubon in colonies numbering tens of thousands. Many other birds are following in their wake.

England and Germany have excellent laws protecting the birds there in summer, only for the Italians to eat during the winter migration. And it is equally useless to have good game and other bird laws in a country like ours, unless they are reinforced in every state by public sentiment against the wanton destruction of bird life for any purpose whatsoever.

This altruism has a solid foundation in economic facts. It is estimated that the farmers of Pennsylvania lost over four millions of dollars one year through the ravages of field mice, because a wholesale slaughtering of owls had been ignorantly encouraged by rewards the year before. Nature adjusts her balances so wisely that we cannot afford to tamper with them.

It is a special pleasure to acknowledge indebtedness to Mr. G. O. Shields. To his efforts, as president of the League of American Sportsmen and as editor of *Recreation*, is due no small measure of the revulsion against ruthless slaughter that has long masqueraded under the disguise of sport.   True sportsmen, worthy of the name, are to be reckoned among the birds' friends, and are doing effective work to help restore those happy hunting grounds which, only a few generations ago, were the envy of the world.

NELTJE BLANCHAN.

LIST OF COLORED PLATES

	FACING PAGE
PASSENGER PIGEON—*Frontispiece*	
PIED-BILLED GREBE	10
LOON	14
BRÜNNICH'S MURRE	22
HERRING GULL	40
COMMON TERN	50
BLACK TERN	58
WILSON'S STORMY PETREL	68
RED-BREASTED MERGANSER	88
MALLARD DUCK	94
BLACK DUCK	98
BALD-PATE DUCK	100
GREEN-WINGED TEAL	104
PIN-TAIL DUCK	110
WOOD DUCK	112
CANVASBACK DUCK	116
GOLDEN-EYE DUCK	122
CANADA GOOSE	138
LEAST BITTERN	158
GREAT BLUE HERON	162
BLACK-CROWNED NIGHT HERON	168
SORA RAIL	180
PURPLE GALLINULE	184
COOT OR MUD HEN	188

List of Colored Plates

	FACING PAGE
Avocet	198
Woodcock	202
Wilson's or Jack Snipe	206
Pectoral Sandpiper or Grass Snipe	212
Least Sandpiper	216
Yellowlegs	224
Bartramian Sandpiper or Upland Plover	230
Golden Plover	240
Semipalmated or Ring Plover	244
Bob White	260
Dusky or Blue Grouse	268
Ruffed Grouse	272
Prairie Hen	278
Prairie Sharp-tailed Grouse	282
Wild Turkey	288
Turkey Vulture	304
Marsh Hawk	312
Red-shouldered Hawk	320
Sparrow Hawk	330
Osprey	334
Saw Whet Owl	342
Screech Owl	344
Great Horned Owl	346
Snowy Owl	350

PART I
WATER BIRDS

TO A WATERFOWL

Whither, 'midst falling dew,
While glow the heavens with the last steps of day,
Far, through their rosy depths, dost thou pursue
    Thy solitary way?

Vainly the fowler's eye
Might mark thy distant flight to do thee wrong,
As, darkly seen against the crimson sky,
    Thy figure floats along.

Seek'st thou the plashy brink
Of weedy lake, or marge of river wide,
Or where the rocking billows rise and sink
    On the chafed ocean side?

There is a Power whose care
Teaches thy way along that pathless coast,
The desert and illimitable air,
    Lone wandering, but not lost.

All day thy wings have fann'd,
At that far height, the cold, thin atmosphere,
Yet stoop not, weary, to the welcome land,
    Though the dark night is near.

And soon that toil shall end;
Soon shalt thou find a summer home and rest,
And scream among thy fellows; reeds shall bend
    Soon o'er thy sheltered nest.

Thou'rt gone, the abyss of heaven
Hath swallowed up thy form; yet, on my heart,
Deeply hath sunk the lesson thou hast given,
    And shall not soon depart.

He who, from zone to zone,
Guides through the boundless sky thy certain flight,
In the long way that I must tread alone,
    Will lead my steps aright.
                  WILLIAM CULLEN BRYANT.

DIVING BIRDS

Grebes
Loons
Auks
Murres
Puffins

DIVING BIRDS

GREBES, LOONS, AUKS, MURRES, PUFFINS
(Order Pygopodes)

The birds of this order, whose Latin name refers to their sitting posture when on land, represent the highest development in the art of swimming and diving, being the nearest lineal descendants of the reptiles, the ancestors of all birds, evolutionists tell us. The American Ornithologists' Union has classified these divers into three distinct families.

Grebes
(Family Podicipidæ)

Grebes, although similar to the loons in general structure and economy, have peculiarly lobed and flattened-out toes connected by webs that are their chief characteristic. In the breeding season several species wear ornamental head-dresses, colored crests or ruffs that disappear in the winter months. Plumage, which is thick, compact, and waterproof, has a smooth, satiny texture, especially on the under parts. Wings, though short, are powerful, and enable the grebes to migrate long distances; but they are not used in swimming under water, as is often asserted. The marvelous rapidity with which grebes dive and swim must be credited to the feet alone. No birds are more thoroughly at home in the water and more helpless on land than they. By keeping only the nostrils above the surface they are able to remain under water a surprising length of time, which trick, with many other clever natatorial feats, have earned for them such titles as "Hell Diver," "Water Witch," and "Spirit Duck." On shore the birds rest upright, or nearly so, owing to the position of their legs, which are

Diving Birds

set far back near the rudimentary tail that serves as a prop to help support the top-heavy, awkward body.

        Holbœll's Grebe
        Horned Grebe
        Pied-billed Grebe or Dabchick

Loons
(Family Urinatoridæ)

Loons, while as famous divers and swimmers as the grebes, are not quite so helpless on land, for they use both bill and wings to assist them over the ground during the nesting season, almost the only time they visit it. They dive literally like a flash, the shot from a rifle reaching the spot sometimes a second after the loon has disappeared into the depths of the lake, where it seems to sink like a mass of lead. It can swim several fathoms under water; also, just below the surface with only its nostrils exposed, and progressing by the help of the feet alone. The sexes are alike. They are large, heavy birds, broad and flat of body, with dark backs spotted with white, and light under parts. Owing to the position of their legs at the back of their bodies, the loons stand in an upright position when on land. The voice is extremely loud, harsh, and penetrating.

        Common Loon
        Black-throated Loon
        Red-throated Loon

Auks, Murres, Puffins
(Family Alcidæ)

Unlike either the grebes or the loons, these diving birds are strictly maritime, passing the greater part of their lives upon the open sea and visiting the coast chiefly to nest. Enormous colonies of them appropriate long stretches of rocky cliffs at the far north at the breeding season, and return to the same spot generation after generation. In spite of their short wings, which are mere flippers, several species fly surprisingly well, although the great auk owed its extinction chiefly to a lack of wing-power. Under water the birds of this family do use their wings to assist in the

Diving Birds

pursuit of fish and other sea-food, which grebes and loons do not, many ornithologists to the contrary notwithstanding. On land the bird moves with a shuffling motion, laboriously and with the underparts often dragging **over** the ground. Agreeing in general aspects, the **birds** of **this** family differ greatly in the form of the bill **in** almost every species. This feature often takes on odd **shapes** during the nesting season, soft parts growing out of the original bill, then hardening into a horny substance, showing numerous ridges and furrows, and sometimes becoming brilliantly colored, only to fade away or drop off bit by bit as winter approaches.

       Puffin or Sea Parrot
       Black Guillemot
       Brünnich's Murre
       Common Murre
       Californian Murre
       Razor-billed Auk
       Dovekie or Sea Dove

THE GREBES, OR LOBE-FOOTED DIVERS
(Family Podicipidæ)

Holbœll's Grebe
(Colymbus holbœllii)

Called also: RED-NECKED GREBE

Length—About 19 inches. Largest of the common grebes.

Male and Female—*In summer:* Upper parts dusky; top of head, small crest, and nape of neck glossy black; throat and cheeks ashy; neck rich chestnut red, changing gradually over the smooth, satiny breast to silvery white or gray dappled under parts; sides also show chestnut tinge. *In winter:* Crests scarcely perceptible; upper parts blackish brown; ashy tint of cheeks and throat replaced by pure white; under parts ashy, the mottling less conspicuous than in summer. Red of neck replaced by variable shades of reddish brown, from quite dark to nearly white. Elongated toes furnished with broad lobes of skin.

Young—Upper parts blackish; neck and sides grayish; throat and under parts silvery white. Head marked with stripes.

Range—Interior of North America from Great Slave Lake to South Carolina and Nebraska. Breeds from Minnesota northward, and migrates southward in winter.

Season—Irregular migrant and winter visitor.

The American, red-necked grebe, a larger variety of the European species, keeps so closely within the lines of family traditions that a description of it might very well serve as a composite portrait of its clan. Six members of this cosmopolitan family, numbering in all about thirty species, are found in North America; the others are distributed over the lakes and rivers of all parts of the world that are neither excessively hot nor cold.

On the border of some reedy pond or sluggish stream, in a floating mass of water-soaked, decaying vegetation that serves as a nest, the red-necked grebe emerges from its dull white egg and instantly takes to water. Cradled on the water, nourished by the wild grain, vegetable matter, small fish, tadpoles, and insects the water supplies, sleeping while afloat, diving to pursue fish and escape danger, spending, in fact, its entire time in or about the water, the grebe appears to be more truly a water-fowl than any of our birds. On land, where it almost never ventures, it is ungainly and uncomfortable; in the water it is marvelously graceful and expert at swimming and diving; quick as a flash to drop out of sight, like a mass of lead, when danger threatens, and clever enough to remain under water while striking out for a safe harbor, with only its nostrils exposed above the surface. Ordinarily it makes a leap forward and a plunge head downward with its body in the air for its deep dives. The oily character of its plumage makes it impervious to moisture. Swimming is an art all grebes acquire the day they are hatched, but their more remarkable diving feats are mastered gradually. Far up north, where the nesting is done, one may see a mother bird floating about among the sedges with from two to five fledglings on her back, where they rest from their first natatorial efforts. By a twist of her neck she is able to thrust food down their gaping beaks without losing her balance or theirs. The male bird keeps within call, for grebes are devoted lovers and parents.

It is only in winter that we may meet with these birds in the United States, where their habits undergo slight changes. Here they are quite as apt to be seen near the sea picking up small fish and mollusks in the estuaries, as in the inland ponds and streams. During the migrations they are seen to fly rapidly, in spite of their short wings and heavy bodies, and with their heads and feet stretched so far apart that a grebe resembles nothing more than a flying projectile.

Horned Grebe
(Colymbus auritus)

Called also: DUSKY GREBE; HELL DIVER; SPIRIT DUCK; WATER WITCH; DIPPER

Length—14 inches.

Male and Female—In summer: Prominent yellowish brown crests resembling horns; cheeks chestnut; rest of head with puffy black feathers; back and wings blackish brown with a few

Grebes

whitish feathers in wings; front of neck, upper breast, and sides chestnut; lower breast and underneath, white. *In winter:* Lacking feathered head-dress; upper parts grayish black; under parts silvery white, sometimes washed with gray on the throat and breast. Elongated toes are furnished with broad lobes of skin.

Young—Like adults in winter plumage, but with heads distinctly striped.

Range—From Northern United States northward to fur countries in breeding season; migrating in winter to Gulf States.

Season—Plentiful during migrations in spring and autumn. Winter resident.

The ludicrous-looking head-dress worn by this grebe in the nesting season at the far north has quite disappeared by the time we see it in the United States; and so the bird that only a few months before was conspicuously different from any other, is often confounded with the pied-billed grebe, which accounts for the similarity of their popular names. As the bird flies it is sometimes also mistaken for a duck; but a grebe may always be distinguished by its habit of thrusting its head and feet to the farthest opposite extremes when in the air. No birds are more expert in water than these. When alarmed they sink suddenly like lead, and from the depth to which they appear to go is derived at least one of their many suggestive names. Or, they may leap forward and plunge downward; but in any case they protect themselves by diving rather than by flight, and the maddening cleverness of their disappearance, which can be indefinitely prolonged owing to their habit of swimming with only the nostrils exposed above the surface, makes it simply impossible to locate them again on the lake.

On land, however, the grebes are all but helpless. Standing erect, and keeping their balance by the help of a rudimentary tail, they look almost as uncomfortable as fish out of water, which the evolutionists would have us believe the group of diving birds very nearly are. When the young ones are taken from a nest and placed on land they move with the help of their wings as if crawling on "all fours," very much as a reptile might; and the eggs from which they have just emerged are ellipsoidal—i. e., elongated and with both ends pointed alike, another reptilian characteristic, it is thought. But oology is far from an exact science. As young alligators, for example, crawl on their

PIED BILLED GREBE.

mother's back to rest, so the young grebes may often be seen. With an undertrust from the mother's wing, which answers every purpose of a spring-board, the fledglings are precipitated into the water, and so acquire very early in life the art of diving, which in this family reaches its most perfect development. For a while, however, the young try to escape danger by hiding in the rushes of the lake, stream, or salt-water inlet, rather than by diving.

Grebes are not maritime birds. Their preference is for slow-moving waters, especially at the nesting season, since their nests are floating ones, and their food consists of small fish, mollusks, newts, and grain, such as the motionless inland waters abundantly afford. In winter, when we see the birds near our coasts, they usually feed on small fish alone. Unhappily the plumage of this and other grebes is in demand by milliners and furriers, to supply imaginary wants of unthinking women.

Pied-billed Grebe

(Podilymbus podiceps)

Called also: DABCHICK; DIEDAPPER; LITTLE GREBE; HELL-DIVER; WATER-WITCH; CAROLINA GREBE; DIPPER; DIPCHICK

Length—14 inches. Smallest of the grebes.

Male and Female—*In summer:* Upper parts dusky, grayish brown; wings varied with ashy and white; throat black; upper breast, sides of throat, and sides of body yellowish brown, irregularly and indistinctly mottled or barred with blackish and washed with yellowish brown; lower breast and underneath glossy white. A few bristling feathers on head, but no horns. Bill spotted with dusky and blue (pied-billed) and crossed with a black band. Toes elongated and with broad lobes of skin. *In winter:* Similar to summer plumage, except that throat is white and the black band on bill is lacking.

Young—Like adults in winter. Heads beautifully striped with black, white, and yellowish brown.

Range—British provinces and United States and southward to Brazil, Argentine Republic, including the West Indies and Bermuda, breeding almost throughout its range.

Grebes

Season—Common migrant in spring and fall. Winters from New Jersey and southern Illinois southward.

The most abundant species of the family in the eastern United States, particularly near the Atlantic, the pied-billed grebes are far from being maritime birds notwithstanding. Salt water that finds its way into the fresh-water lagoons of the Gulf States, or the estuaries of our northern rivers, is as briny as they care to taste; and although so commonly met with near the sea, they are still more common in the rivers, lakes, and ponds inland, where tall reeds and sedges line the shores and form their ideal hunting and nesting grounds. The grebes and loons are not edible, nor are they classed as game birds by true sportsmen; nevertheless this bird is often hunted, although the sportsman finds it a wary victim, for there is no bird in the world more difficult to shoot than a "water-witch." One instant it will be swimming around the lake apparently unconcerned about the intruder; the next instant, and before aim can be taken, it will have dropped to unknown depths, but presumably to the infernal regions, the sportsman thinks, as he rests meditatively upon his gun, waiting for the grebe to reappear in the neighborhood, which it never dreams of doing. It will swim swiftly under water to a safe distance from danger; then, by keeping only its nostrils exposed to the air, will float along just under the surface and leave its would-be assassin completely mystified as to its whereabouts —a trick the very fledglings practice. It is amazing how long a grebe can remain submerged. In pursuing fish, which form its staple diet; in diving to escape danger, to feed, to loosen water-weeds for the construction of its nest, among its other concerns below the surface, it has been missed under water for five minutes, and not at all short of breath on its return above at the end of that time. Fresh-water mollusks, newts, winged insects, vegetable matter, including seeds of wild grain and some grasses, vary the bird's fish diet.

Ungainly and ill at ease on land, in fact, almost helpless there, a grebe rarely ventures out of the water either to sleep or to nest. The young rest on their mother's back after their first swimming lessons that are begun the hour they are hatched; but they quickly become wonderfully expert and independent of everything except water: that is their proper element. Nevertheless they can fly with speed and grace, though with much working of their short wings and stretching of their short bodies, from which their heads project as far as may be at one end and their great lobed feet at the other.

The nest of all grebes is an odd affair, one of the curiosities of bird architecture. A few blades of "saw grass" may or may not serve as anchor to the floating mass of water-weeds pulled from the bottom of the lake and held together by mud and moss. The structure resembles nothing so much as a mud pancake rising two or three inches above the water, though, like an iceberg, only about one-eighth of it shows above the surface. A grebe's nest is often two or three feet in depth. In a shallow depression, from four to ten, though usually five, soiled, brownish-white eggs are laid, and concealed by a mass of wet muck whenever the mother leaves her incubating duties. At night she sits on the nest, and for some hours each day; but at other times the water-soaked, muck-covered cradle, with the help of the sun, steams the contents into life.

THE LOONS
(Family urinatoridæ)

Loon
(Urinator imber)

Called also: GREAT NORTHERN DIVER; COMMON LOON; LOOM

Length—31 to 36 inches.

Male and Female—In summer: Upper parts glossy black, showing iridescent violet and green tints. Back and wings spotted and barred with white; white spaces on the neck marking off black bands, and sides of breast streaked with white. Breast and underneath white. Bill stout, straight, sharply pointed, and yellowish green. Legs, which are placed at rear of body, are short, buried and feathered to heel joint. Tail short, but well formed. Feet black and webbed. *In winter and immature specimens:* Upper parts blackish and feathers margined with grayish, not spotted with white. Underneath white; throat sometimes has grayish wash.

Range—Northern part of northern hemisphere. In North America breeds from the Northern United States to Arctic Circle, and winters from the southern limit of its breeding range to the Gulf of Mexico.

Season—A wandering winter resident. Most common in the migrations from September to May.

This largest and handsomest of the diving birds, as it is the most disagreeably voiced, comes down to our latitude in winter, when its favorite inland lakes at the north begin to freeze over and the fish to fail, and wanders about far from the haunts of men along the seacoast or by the fresh waterways. Cautious, shy, fond of solitude, it shifts about from place to place discouraging our acquaintance. By the time it reaches the United States—for the majority nest farther north—it has exchanged its rich, velvety black and white wedding garment for a more dingy suit, in which the immature specimens are also dressed. With

LOON.

strong, direct flight small companies of loons may be seen high overhead migrating southward to escape the ice that locks up their food; or a solitary bird, some fine morning in September, may cause us to look up to where a long-drawn, melancholy, uncanny scream seems to rend the very clouds. Nuttall speaks of the "sad and wolfish call which like a dismal echo seems slowly to invade the ear, and rising as it proceeds, dies away in the air. This boding sound to mariners, supposed to be indicative of a storm, may be heard sometimes two or three miles when the bird itself is invisible, or reduced almost to a speck in the distance." But the loon has also a soft and rather pleasing cry, to which doubtless Longfellow referred in his "Birds of Passage," when he wrote of

> . . . "The loon that *laughs* and flies
> Down to those reflected skies."

. Not so aquatic as the grebes, perhaps the loons are quite as remarkable divers and swimmers. The cartridge of the modern breech-loader gives no warning of a coming shot, as the old-fashioned flint-lock did; nevertheless, the loon, which is therefore literally quicker than a flash at diving, disappears nine times out of ten before the shot reaches the spot where the bird had been floating with apparent unconcern only a second before. As its flesh is dark, tough, and unpalatable, the sportsman loses nothing of value except his temper. Sometimes young loons are eaten in camps where better meat is scarce, and are even offered in large city markets where it isn't.

In spring when the ice has broken up, a pair of loons retire to the shores of some lonely inland lake or river, and here on the ground they build a rude nest in a slight depression near enough to the water to glide off into it without touching their feet to the sand. In June two grayish olive-brown eggs, spotted with umber brown, are hatched. The young are frequently seen on land as they go waddling about from pond to pond. After the nesting season the parents separate and undergo a moult which sometimes leaves so few feathers on their bodies that they are unable to rise in the air. When on land they are at any time almost helpless and exceedingly awkward, using their wings and bill to assist their clumsy feet.

Loons

The Black-throated Loon *(Urinator arcticus)*, a more northern species than the preceding, reaches only the Canadian border of the United States in winter. It may be distinguished from the common loon by its smaller size, twenty-seven inches, and by its gray feathers on the top of the head and the nape of the neck, though in winter plumage even this slight difference of feathers is lacking.

Red-throated Loon

(Urinator lumme)

Called also : SPRAT LOON; RED-THROATED DIVER; COBBLE

Length—25 inches.

Male and Female—In summer: Crown and upper parts dull brownish black, with a greenish wash and profusely marked with white oval spots and streaks. Underneath white. Bluish gray on forehead, chin, upper throat, and sides of head. A triangular mark of chestnut red on fore neck. Bill black. Tail narrowly tipped with white. *In winter and immature specimens:* Similar to the common loon in winter, except that the back is spotted with white.

Range—Throughout northern parts of northern hemisphere; migrating southward in winter nearly across the United States.

Season—Winter visitor or resident.

It is not an easy matter at a little distance to distinguish this loon from the great northern diver, for the young of the year, which are most abundant migrants in the United States, lack the chestnut-red triangle on the throat, which is the bird's chief mark of identification. Its smaller size is apparent only at close range. In habits these loons are almost identical; and although their name, used metaphorically, has come to imply a simpleton or crazy fellow, no one who has studied them, and certainly no one who has ever tried to shoot one, can call them stupid. It is only on land, where they are almost never seen, that they even look so.

Audubon found the red-throated loons nesting on the coast of Labrador, near small fresh-water lakes, in June. The young are able to fly by August, and in September can join the older migrants in their southern flight. In England these loons follow the movements of the sprats, on which they feed; hence one of their common names by which our Canadian cousins often call them. Fishermen sometimes bring one of these divers that has been gorging on the imprisoned fish, to shore in their nets. For a fuller account of the bird's habits, see the common loon.

AUKS, MURRES, PUFFINS
(Family Alcidæ)

Puffin
(Fratercula arctica)

Called also: SEA PARROT; COULTERNEB; MASKING PUFFIN

Length—13 inches.

Male and Female—Upper parts blackish; browner on the head and front of neck. Sides of the head and throat white; sometimes grayish. Nape of neck has narrow grayish collar. Breast and underneath white. Feet less broadly webbed than a loon's. Bill heavy and resembling a parrot's. In nesting season bill assumes odd shapes, showing ridges and furrows, an outgrowth of soft parts that have hardened and taken on bright tints. A horny spine over eye. Colored rosette at corner of mouth.

Range—Coasts and islands of the North Atlantic, nesting on the North American coast from the Bay of Fundy northward. South in winter to Long Island, and casually beyond.

Season—Winter visitor.

Few Americans have seen this curious-looking bird outside the glass cases of museums; nevertheless numbers of them straggle down the Atlantic coast as far as Long Island every winter, from the countless myriads that nest in the rocky cliffs around the Gulf of St. Lawrence and the Bay of Fundy. Unlike either grebes or loons, puffins are gregarious, especially at the nesting season. In April great numbers begin to assemble in localities to which they return year after year, and select crevices in the rocks or burrow deep holes like a rabbit, to receive the solitary egg that is the object of so much solicitude two months later. Both male and female work at excavating the tunnel and at feeding their one offspring, which has an appetite for fish and other sea-food large enough for a more numerous family. By the end of August the entire colony breaks up and follows the exodus of fish, completely deserting their nesting grounds, where any young ones that may be hatched late are left to be preyed upon by hawks and ravens. "Notwithstanding this apparent neglect of their young at this time, when every other instinct is merged in the desire and necessity of migration," wrote Nuttall, "no bird is more attentive to them in general, since they will suffer themselves to be taken by the hand and use every endeavor to save and screen their young, biting not only their antagonist, but, when laid hold of by the wings, inflicting bites on themselves, as if actuated by the agonies of despair; and when released, instead of flying away, they hurry again into the burrow." A hand thrust in after one may drag the angry parent, that has fastened its beak upon a finger, to the mouth of the tunnel; but a certain fisherman off the coast of Nova Scotia, who lost a piece of solid flesh in this experiment, now gives advice freely against it.

The beak that is able to inflict so serious an injury is this bird's chief characteristic. It looks as if it had been bought at a toyshop for some reveller in masquerade; but the puffin wears it only when engaged in the most serious business of life, for it is the wedding garment donned by both contracting parties. It is about as long as the head, as high as it is long, having flat sides that show numerous ridges or furrows from the fact that each represents new growth of soft matter that finally hardens into horn as the nesting season approaches, only to disappear bit by bit until nine pieces have been moulted or shed, very much as a deer casts its antlers. The white pelican drops its "centreboard" in a similar manner. In the puffins there is also a moult of the excrescenses upon the eyelids, and a shrivelling of the colored rosette at the corner of the mouth, peculiarities first scientifically noted by L. Bereau about twenty years ago. The change of plumage after moult is scarcely perceptible.

On land the bird walks upright, awkwardly shuffling along on the full length of its legs and feet. It is an accomplished swimmer and diver, like the grebes and loons, although, unlike them, it uses its wings under water. When a strong gale is blowing off the coast, the puffins seek shelter in the crevices of the rocks or their tunnels in the sand; but some that were overtaken by it on the open sea, unable to weather it, are sometimes found washed ashore dead after a violent storm. Mr. Brewster,

Auks, Murres, Puffins

who made a special study of these birds in the Gulf of St. Lawrence, writes: "The first report of our guns brought dozens tumbling from their nests. Their manner of descending from the higher portions of the cliff was peculiar. Launching into the air with heads depressed and wings held stiffly at a sharp angle above their backs, they would shoot down like meteors, checking their speed by an upward turn just before reaching the water. In a few minutes scores had collected about us. They were perfectly silent and very tame, passing and repassing over and by us, often coming within ten or fifteen yards. On such occasions their flight has a curious resemblance to that of a woodcock, but when coming in from the fishing grounds they skim close to the waves and the wings are moved more in the manner of those of a duck."

Black Guillemot
(Cepphus grylle)

Called also: SEA PIGEON

Length—13 inches.

Male and Female—In summer: Prevailing color sooty black, with greenish tints above and lighter below. Large white patch on upper wings, and white ends of wing feathers, leave a black bar across the wings, sometimes apparently, though not really, absent; wing linings white. Bill and claws black; mouth and feet vermilion or pinkish. *In winter:* Wings and tail black, with white patch on wings; back, hind neck, and head black or gray variegated with white. Under parts white.

Young—Upper parts like adults in winter, except that the under parts are mottled with black. Nestlings are covered with blackish-brown down. Feet and legs blackish.

Range—Breeds from Maine to Newfoundland and beyond; migrates south in winter, regularly to Cape Cod, more rarely to Long Island, and casually as far as Philadelphia.

Small companies of sea pigeons, made up of two or three pairs that keep well together, may be seen almost grazing along the surface of the sea off our northern States and the Canadian coast, following a straight line at the base of the cliffs while keeping a sharp lookout for the small fish, shrimps, baby crabs, and marine insects they pick up on the way. Suddenly one of the birds dives after a fish, pursues, overtakes, and swallows it, then rejoins its mate with little loss of time; for these sea pigeons use their wings under water as well as above it, and so are able to reappear above the surface at surprising distances from the point where they went down. They are truly marine birds; never met with inland, and rarely on the shore itself, except at the nesting season. Large companies nest in the crevices and fissures of cliffs and rocky promontories, heaping up little piles of pebbles that act as drains for rainwater or melting snow under the eggs. Incubation takes place in June or July, according to the latitude. Two or three sea-green or whitish eggs, irregularly spotted and blotched with blackish brown, and with purplish shell-markings, make up a clutch.

In the diary kept on the *Jeannette*, De Long recorded meeting with black guillemots in latitude 73°, swimming about in the open spaces between the ice-floes early in May; and Greely ate their eggs off the shores of Northern Greenland in July. Both explorers mentioned the presence of fox tracks in the neighborhood of the guillemots, proving that this arch enemy pursues them even into the desolation of the Arctic Circle. One of the first lessons taught the young birds is to hurl themselves from the jutting rocks to escape the fox that is forever threatening their lives in the eyries, and to dive into the sea that protects and feeds them.

Brünnich's Murre

(Uria lomvia)

Called also : BRÜNNICH'S GUILLEMOT; ARRIE; EGG BIRD; PENGUIN; FOOLISH GUILLEMOT

Length—16.50 inches.
Male and Female—Sooty black above, brownest on front of neck. Breast and underneath, white. White tips to secondaries form an obscure band. Greenish base to the upper half of bill, which is rounded outward over the lower half. Bill short, stout, wide, and deep.
Range—Coasts and islands of the North Atlantic and eastern Arctic Oceans. South to the lakes of Northern New York and the coast of New Jersey. Nests from the Gulf of St. Lawrence northward.
Season—Winter visitor in United States.

"The bird cliffs on Arveprins Island (Northern Greenland) deserve a passing notice, not for Arctic travellers, but for the general reader," writes General Greely in "Three Years of Arctic Service."

"For over a thousand feet out of the sea these cliffs rise perpendicularly, broken only by narrow ledges, in general inaccessible to man or other enemy, which afford certain kinds of sea fowl secure and convenient breeding places. On the face of these sea-ledges of Arveprins Island, Brünnich's guillemots, or loons, (*sic*) gather in the breeding season, not by thousands, but by tens of thousands. Each lays but a single gray egg, speckled with brown; yet so numerous are the birds, that every available spot is covered with eggs. The surprising part is that each bird knows its own egg, although there is no nest and it rests on the bare rock. Occasional quarrels over an egg generally result in a score of others being rolled into the sea.

"The clumsy, short-winged birds fall an easy prey to the sportsman, provided the cliffs are not too high, but many fall on lower inaccessible ledges, and so uselessly perish. A single shot brings out thousands on the wing, and the unpleasant cackling, which is continuous when undisturbed, becomes a deafening clamor when they are hunted.

"The eggs are very palatable. The flesh is excellent—to my taste the best flavored of any Arctic sea fowl; but, to avoid the slightly train-oil taste, it is necessary to keep the bird to ripen, and to carefully skin it before cooking." Later on, the starving survivors in the camp near Cape Sabine owed the prolonging of their wretched existence from day to day largely to these very birds.

When these murres come down from the far north to visit us in winter they keep so well out from land that none of our ornithologists seem to have made a very close study of them. Like other birds of the order to which they belong, they dive suddenly out of sight when approached, and by the help of wings and feet swim under water for incredible distances.

. . . .

The Common Murre or Guillemot *(Uria troile)*, so called, is certainly less common in the United States than the preceding species. Massachusetts appears to be its southern limit. In winter, when we see it here, it can be distinguished from

BRUNNICHS MURRE.
⅓ Life-size.

Brünnich's murre only by its bill, which is half an inch longer. Some specimens show a white ring or "eye-glass" around the eye and a white stripe behind it; but doubt exists as to whether such specimens are not a separate species. Much study has still to be given to this group of birds before the differences of opinion held by the leading ornithologists concerning them will be settled satisfactorily to all. The habits of the three murres mentioned here are identical so far as they are known. Penguin and foolish guillemot are titles sometimes given to the common murre; but to add to popular confusion, they are just as frequently applied to Brünnich's murre.

The Californian murre, the Western representative of these species, differs from them neither in plumage nor habits, it is said. It breeds abundantly from Behring's Sea to California, and the natives of Alaska depend upon its eggs for food. They were among the first dainties sold to the Klondike miners.

Razor-billed Auk
(Alca torda)

Called also : TINKER

Length—16.50 inches.

Male and Female—In summer: Upper parts sooty black; browner on fore neck. A conspicuous white line from eye to bill; breast, narrow line on wing, wing-linings, and underneath, white. Bill, which is about as long as head, and black, has horny shield on tip and is crossed by sunken white band. Tail upturned. *In winter:* Similar to summer plumage, except that it is duller and the sides and front of neck are white. Bill lacks horny shield. White line on bill, sometimes lacking on winter birds and always on immature specimens.

Range—"Coasts and islands of the North Atlantic; south in winter on the North American coast, casually to North Carolina. Breeding from Eastern Maine northward." A. O. U.

Season—Winter visitor.

Audubon, who followed these birds to their nesting haunts in Labrador and the Bay of Fundy, found the bodies of thousands strewn on the shores, where, after their eggs had been taken by boat loads for food, and the fine, warm feathers of their breasts had been torn off for clothing, they were left to decay. In Nova Scotia he met three men who made a business of egg-hunting. They began operations by trampling on all the eggs they found laid, relying on the well-known habit of the auk and its relatives that lay but a single egg, to replace it should it be destroyed. Thus they made sure of fresh eggs only. In the course of six weeks they had collected thirty thousand dozen, worth about two thousand dollars. As this wholesale destruction of our gregarious marine birds has been going on for a century at least, is it not surprising that they are not all extinct, like the great auk?

Without wings to help them escape from the voyagers and fishermen who pursued them on sea and ashore, the great auks, that in Nuttall's day were still breeding in enormous colonies in Greenland, dwindled to a single specimen "found dead in the vicinity of St. Augustine, Labrador, in November, 1870," which, although in poor condition, was sold for two hundred dollars to a European buyer. The Smithsonian Institution, the Philadelphia Academy, Cambridge Museum, and Vassar College own one specimen each, the only ones in this country, so far as known.

The moral from the story of the great auk that the razor-billed species and its short-winged relatives should take to heart, obviously, is to keep their wings from degenerating into useless appendages, by constant exercise. They certainly are strong flyers in their present evolutionary stage, and, by constantly flapping their stiffened wings just above the level of the sea, are usually able to escape pursuit, if not in the air then by diving through the crest of a wave and still using their wings as a fish would its fins, to assist their flight under water. Though they move awkwardly on land, so awkwardly as to suggest the possible derivation of the adverb from their name, they still move rapidly enough to escape with their life in a fair race. When cornered, the hand that attempts to seize them receives a bite that sometimes takes the flesh from the bone—such a bite as the sea parrot gives.

In the nesting grounds, where enormous numbers of these razor-billed auks have congregated from times unknown, the females may be seen crouching along the eggs, not across them, in long, seriate ranks, where tier after tier of cliffs rise from the water's edge to several hundred feet above the sea. Where there is no attempt at a nest, and each buffy and brown speckled egg looks just like the thousands of others lying loosely about in the rocky crevices, it is amazing how each bird can tell its own. The male birds are kept busy during incubation bringing small fish in their bills to their sitting mates or relieving them on the eggs while the females go a-fishing. For a short time only the young birds are fed by regurgitation; then small fish are laid before them for them to help themselves, and presently they go tumbling off the jutting rocks into the sea to dive and hunt independently. Particularly at the nesting season these razor-bills utter a peculiar grunt or groan; but the stragglers from the great flocks that reach our coast in winter are almost silent.

Dovekie
(Alle alle)

Called also : SEA DOVE; LITTLE AUK; PIGEON DIVER; GREENLAND DOVE; ICE BIRD

Length—8.50 inches.

Male and Female—In summer : Upper parts, including head and neck all around, glossy black, shoulders and other wing feathers tipped with white and forming two distinct patches. Lower breast and underneath white. A few white touches about eyes. Wings long for this family. Body squat, owing to small, weak feet. Wing linings dusky. *In winter:* Resembling summer plumage, except that the black upper parts become sooty and the white of lower breast extends upward to the bill, almost encircling the neck. Sometimes the white parts are washed with grayish and the birds have gray collar on nape.

Young—Like adults in winter, but their upper parts are duller.

Range—From the farthest north in the Atlantic and Arctic oceans, south to Long Island, and occasionally so far as Virginia.

Season—Winter visitor.

In the chapter entitled "The End—by Death and by Rescue," in his "Three Years of Arctic Service," General Greely, after telling how the wretched men at Cape Sabine were reduced to eating their sealskin boots and were apparently in the last extremity, goes on to describe how Long, one of the hunters of the expedition, one awful day succeeded in shooting four of these little dovekies, two king-ducks, and a large guillemot. But the current swept away all the birds except one dovekie! "I ordered the dovekie to be issued to the hunters who can barely walk," writes the starving commander; "but . . . one man begged with tears for his twelfth, which was given him with everybody's contempt." When the twelfth part of a little bird that a man can easily cover with his hand causes a scene like this, can the imagination picture the harrowing misery of the actual situation?

And yet where man and nearly every other living creature perishes, the little auk pursues its happy way, floating about in the open water, left even in that Arctic desolation by the drifting ice floes, and diving into its icy depths after the shrimps that Greely's party collected at such frightful cost.

Far within the Arctic Circle great colonies nest after the fashion of their tribe, in the jutting cliffs that overhang the sea. One pale, bluish-white egg, laid on the bare rock, is all that nature requires of these birds to carry on the species, whose chief protection lies in their being able to live beyond the reach of men, to escape pursuit by diving and rapid swimming under water, and to fly in the teeth of a gale that would mean death to a puffin. With so many means of self-preservation at their disposal, there is no need of a large family to keep up the balance that nature adjusts.

These neat little birds, whose form alone suggests a dove, are by no means the lackadaisical creatures their name seems to imply. They are self-reliant, for they are chiefly solitary birds that straggle down our coast in winter. They are wonderfully quick of motion in their chosen element, and although they have a peculiar fashion of splashing along the surface of the water, as if unable to fly, they certainly are in no immediate danger of becoming extinct from the loss of wings through disuse, like the great auk. A little sea dove that once flew across the bow of an ocean steamer in the North Atlantic in an instant became a mere speck in the bleak wintry sky, and the next second vanished utterly.

LONG-WINGED SWIMMERS

Jaegers
Gulls
Terns

LONG-WINGED SWIMMERS

JAEGERS, GULLS, TERNS
(Order Longipennes)

Birds of this order may be recognized among the webbed-footed birds by their long, pointed wings that reach beyond the base of the tail, and in many instances beyond the end of it. They do not hold themselves erect when ashore, as the grebes, loons, and auks do, but are able to keep a horizontal position because their legs are placed nearly, if not perfectly, under the centre of equilibrium. Bills of variable forms, sharply pointed and frequently hooked like a hawk's. Four toes, three of them in front, flat and webbed; a very small rudimentary great toe (hallux) elevated above the foot at the back.

Jaegers and Skuas
(Family Stercorariidæ)

End of upper half of bill is more or less swollen and rounded over the tip of lower mandible. Upper parts of plumage, and sometimes all, sooty, brownish black, frequently with irregular bars. Middle feathers of square tail are longest. The name jaeger, meaning hunter, might be freely translated into pirate; for these creatures of spirited, vigorous flight delight in pursuing smaller gulls and terns to rob them of their fish, like the marine birds of prey that they are. Jaegers and skuas are birds of the seacoast or large bodies of inland water, and wander extensively except at the nesting season in the far North.

      Parasitic Jaeger.
      Pomarine Jaeger.
      Long-tailed Jaeger.

Long-winged Swimmers

Gulls and Terns
(Family Laridæ)

The Gulls
(Subfamily Larinæ)

Bills of moderate length, the upper mandible not swollen at the tip like the jaegers, but curved over the end of the lower mandible. Wings long, broad, strong and pointed, though their flight is less graceful than a tern's. Tail feathers usually of about equal length. Sexes alike, but the plumage, in which white, brown, black, and pearl-blue predominate, varies greatly with age and season. In flight the bill points forward, not downward like a tern's. Gulls pick their food from the surface of the sea or shore, whereas terns plunge for theirs. Gulls are the better swimmers, and pass the greater part of their lives at sea, coming to shore chiefly to nest in large colonies.

    Kittiwake Gull.
    Glaucous Gull, or Burgomaster.
    Iceland Gull.
    Great Black-backed Gull.
    Herring Gull.
    Ring-billed Gull.
    Laughing Gull.
    Bonaparte's Gull.

Terns
(Subfamily Sterinæ)

Small birds of the coast rather than the open sea. Bill straight, not hooked, and sharply pointed. Outer tail feathers longer than the middle ones; tails usually very deeply forked. Legs placed farther back than a gull's, and form of body more slender and trim. Great length and sharpness of wing give a dash to their flight that the gull's lacks. Bill held point downward, like a mosquito's, when tern is searching for food. Plumage scarcely differs in the sexes, but it varies greatly with the season and age. Usually the top of head is black; in the rest of the plumage pearl grays, browns, and white predominate. Tails generally long and forked, so that in aspect, as in flight, the birds suggest their name of sea swallow.

        Marsh Tern.
        Royal Tern.
        Wilson's Tern, or Common Tern.
        Roseate Tern.
        Arctic Tern.
        Least Tern.
        Black Tern.

Skimmers
(Family Rynchopidæ)

Only one species of skimmer inhabits the Western Hemisphere. These birds have extraordinary bills, thin, and resembling the blade of a knife, with lower half much longer than the upper mandible, and used to skim food from the surface of the water and to open shells. Wings exceedingly long; flight more measured and sweeping than a tern's.

        Black Skimmer, or Scissor Bill.

JAEGERS AND SKUAS
(Family *Stercorariidæ*)

Parasitic Jaeger
(*Stercorarius parasiticus*)

Called also: MAN-OF-WAR; ARCTIC JAEGER; RICHARD-SON'S JAEGER; TEASER

Length—17.20 inches.

Male and Female—Light stage: Top of head and cheeks brown, nearly black; back, wings, and tail slaty brown, which becomes reddish brown on sides of breast and flanks. Sides of head, back of neck, and sometimes entire neck and throat yellowish. Under parts white. Wings moderately long, strong and pointed. Middle feathers of tail longest. Black tip of upper half of slate-colored bill is swollen and rounded over end of lower mandible like a hawk's. Feet black. *Dark stage:* Plumage dark slaty brown, darker on top of head, very slightly lighter on under parts. Immature specimens, which seem to be most abundant off our coasts, show sooty slate plumage; bordered, tipped, or barred with buffy, rufous, or brownish black, giving the bird a mottled appearance. Plumage extremely variable with age and season.

Range—Nests in Barren Grounds, Greenland, and other high northern districts; migrates southward along the Atlantic and Pacific coasts and through the Great Lakes, wintering from New York, California, and the Middle States to Brazil.

Season—October to June. Winter visitor.

This dusky pirate, strong of wing and marvelously skilful and alert in its flight, uses its superior powers chiefly to harass and prey upon smaller birds. Lashing the air with its long tail, and with wide wing stretchings and powerful strokes, the jaeger comes bearing down on a kittiwake gull that holds a dripping fish ready for a contemplated dinner. To dart away from its tormentor, that darts, too, even more suddenly; to outrace the jaeger, although freighted with the fish, are tried resorts that the little gull must finally despair of when the inevitable moment arrives that the coveted fish has to be dropped for the pirate to snatch up and bear away in triumph.

Other gulls than the kittiwake suffer from this ocean prowler; their young and eggs are eaten, their food is taken out of their very mouths. As they live so largely on the results of other birds' efforts, the jaegers deserve to be branded as parasites, which all the group are. Indeed, these birds that the English call skuas, differ very little, if any, in habits. While all spend the summer far north, the parasitic jaeger has really less claim to the title of Arctic jaeger than either the pomarine or the long-tailed species, which go within the Arctic Circle to nest. On an open moor or tundra, in a slight depression of the ground, a rude nest is scantily lined with grass, moss, or leaves. Sometimes this nest is near the margin of the sea or lake, sometimes on an ocean island and laid among the rocks. It contains from two to four—usually two—light olive-brown eggs that are frequently tinged with greenish and scrawled over with chocolate markings most plentiful at the larger end, where they may run together and form a blotch.

By the end of September the jaegers begin their southerly migration, reaching Long Island in October, regularly, and quite as regularly leaving early in June. During the winter they play the rôle of sea scavengers when they are not robbing the gulls, that will actually disgorge a meal already safely stowed away rather than submit to the harassing, petty tortures of these pirates. Jaegers constantly pick up carrion and other rubbish cast up by the sea or thrown overboard from a passing ship, for nothing in the line of food, however putrid it may be, seems to miss the mark of their rapacious appetites, as their Latin name, *stercorarius*, a scavenger, indicates. On land they always seek choicer food, garnered by their own effort—berries, insects, eggs, little birds, and mammals.

The best trait the jaegers have is their uncommon courage. Nothing that attacks their home or young is too large or fierce for them to dash at fearlessly; and by persistent teasing and harassing, for the want of formidable weapons of defense, they will eventually get the better of their antagonist, though it be a sea eagle.

. . . .

The Pomarine Jaeger—a contraction of pomatorhine, meaning flap-nosed—*(Stercorarius pomarinus)* may be distinguished from the parasitic jaeger by its larger size, twenty-two inches; by the rounded ends of its central tail feathers, which project about three inches beyond the others; and finally by its darker, almost black, upper parts, although the plumage during the dark and the light phases of these birds is so nearly the same that when seen on the wing it is impossible to tell one species from another. Professor Newton, of Cambridge University, has noted that the long, central tail feathers of the pomarine jaeger have their shafts twisted toward the tip, so that in flight the lower surfaces of their webs are pressed together vertically, giving the bird the appearance of having a disk attached to its tail. This species is also called the pomarine hawk-gull.

It is not known whether the Long-tailed Jaeger, or Buffon's Skua, as they call it in England *(Stercorarius longicaudus)*, undergoes the remarkable changes of plumage that its relatives indulge in or not, for its range is more northerly than that of any of the jaegers, and when it migrates south of the Arctic Circle, to our coasts, it is wearing feathers most confusingly like those of the parasitic jaeger in its light phase. Indeed, the young of these two species cannot be distinguished except by measuring their bills, when it is found that the long-tailed jaeger has the shorter bill.

The distinguishing mark of the adults of this species is the length of the central tail feathers, narrow and pointed, that project about seven inches beyond the others; but immature specimens lack even this mark. The description of the habits of the parasitic jaeger applies equally well to all of the three freebooters mentioned.

GULLS AND TERNS
(Family Laridæ)

Gulls
(Subfamily Larinæ)

Kittiwake
(Rissa tridactyla)

Length—16 inches.

Male and Female—*In summer:* Deep pearl gray mantle over back and wings. Head, neck, tail, and under parts pure white. Ends of outer wing feathers—primaries—black, tipped with white. Tips of tail quills black. Hind toe very small, a mere knob, and without a claw. Bill light yellow. Feet webbed and black. *In winter:* Similar to summer plumage, but that the mantle is a darker gray and extends to back of neck. Dark spot about the eye.

Range—Arctic regions, south in eastern North America in winter to the Great Lakes and the coast of Virginia. Breeds from Magdalen Islands northward.

Season—Autumn and winter visitor in the Middle States. Common north of them all winter.

It is the larger herring gull that we see in such numbers in our harbors and following in the path of vessels along our coast; but the watchful eye may often pick out a few kittiwakes in the loose flocks, and north of Rhode Island meet with a company of them apart from others of their kin. Skimming gracefully along the surface of the water, soaring, floating in mid-air, swooping for a morsel in the trough of the waves, then with a few strong wing strokes rejoining their fellows as they play at cross-tag in the sky, the gulls fascinate the eyes and beguile many a weary hour at sea.

Along the shores of the Arctic Ocean, on the craggy cliffs of Greenland, and beyond, large colonies of kittiwakes nest on the ledges of rock barely scattered over with grass, moss, and seaweed to form a rude nest, or else directly on the sand in the midst of a little heap of "drift" cast high up on the beach. Three or four eggs, varying from buffy to grayish brown, and marked with chocolate, are often taken from a nest by the natives, who, with the jaegers and the sea eagles that also devour the young, are the kittiwakes' worst enemies. Fearlessly breasting a gale on the open ocean, sleeping with head under wing while riding the waves, the gull is far more at home at sea than ashore, and soon leaves the nest to begin its roving life at sea.

Their service to man, aside from the gulls' æsthetic value, is in devouring refuse that would otherwise wash ashore and pollute the air. This is the gull that the jaegers, those dusky pirates of the high seas, most persecute by taking away its fish and other food to save themselves the trouble of hunting in the legitimate way.

Glaucous Gull
(Larus glaucus)

Called also: BURGOMASTER ; ICE GULL.

Length—28 to 32 inches.

Male and Female—In summer: Mantle over wings and back, light pearl gray ; all other parts pure white. Large, strong, wide bill which is chrome yellow, with orange red spot at the angle. Legs and feet pale pink or yellowish pink. *In winter:* Light streaks of pale brownish gray on head and back of neck ; otherwise plumage same as summer. Immature birds are wholly white, with flesh-colored bills having black tips. Females are smaller than males.

Range—Northern and Arctic Oceans around the world; in North America from Long Island and the Great Lakes in winter, to Labrador and northward in the nesting season.

Season—Irregular winter visitor.

This very large gull, whose protective coloring indicates that the snow and ice of the circum-polar regions are its habitual surroundings, occasionally struggles down our coasts and to the Great Lakes in loose flocks in winter, but leaves none too good a character behind it on its departure in the early spring. General Greely met enormous numbers of burgomasters in the dreary desolation of ice at the far north ; and Frederick Schwatka tells of great nesting colonies in the cliffs overhanging the upper waters of the Yukon, where the sound of the rushing torrent was drowned by their harsh uproar as they wheeled about in dense clouds high above his head. The nest, which is a very slight affair of seaweed, moss, or grass, contains two or three stone-colored eggs, although sometimes pale olive-brown ones are found, spotted and marked with chocolate and ashy gray. Many nests are also made directly on the ground.

What is reprehensible in this bird's habits is its tyranny over smaller, weaker gulls and other birds that it hunts down like a pirate to rob of their food while they carry it across the waves or to their nest, where the villain still pursues them and devours their young. Quite in keeping with such unholiness is the burgomaster's harsh cry, variously written *kuk-lak'* and *cut-leek'*, that it raises incessantly when hungry, and that therefore must be particularly unpleasant to the kittiwakes, guillemots, and other conspicuous victims of its rapacious appetite. When its hunger is appeased, however, by fish, small birds, crow-berries, carrion, and morsels floating on the sea, this gull is said to be inactive and silent; and certainly the starving hunters in the Greely expedition found it sadly shy.

The Iceland Gull *(Larus leucopterus)* looks like a small edition of the burgomaster, its length being about twenty-five inches; but its plumage is identical with that of the larger bird.

Great Black-backed Gull
(Larus marinus)

Called also : SADDLE-BACK ; COBB; COFFIN CARRIER

Length—29 to 30 inches.

Male and Female—In summer: Mantle over back and wings dark slaty brown, almost black ; wing feathers tipped with white; rest of plumage white. Bill yellow, red at the angle. Feet and legs pinkish. *In winter:* Similar to summer dress except that the white head and neck are streaked with grayish. Immature birds are mottled brown and white, the perfect plumage described above not being attained until the fourth year.

Range—Coasts of North Atlantic. Nests from Nova Scotia northward. Migrates in winter sometimes to South Carolina and Virginia, but regularly to Long Island and the Great Lakes. *Season*—September to April.

The black-back shares the distinction with the burgomaster of being not only one of the largest, most powerful representatives of its family, but one of the most tyrannical and greedy. So optimistic a bird-lover as Audubon said that it is as much the tyrant of the sea fowl as the eagle is of the land birds. Like the eagle again, it is exceedingly shy of men and inaccessible. "By far the wariest bird that I have ever met," writes Brewster. This same careful observer reports that he noted four distinct cries: "a braying *Ha-ha-ha*, a deep *keow, keow*, a short barking note, and a long-drawn groan, very loud and decidedly impressive," when he studied it in the island of Anticosti.

Soaring high in the air in great spirals, with majestic grace and power, the saddle-back still keeps a watchful eye on what is passing in the world below, and, quick as a hawk, will come swooping down to pounce upon some smaller gull or other bird that has just secured a fish by patient toil, to suck the eggs in a nest left for the moment unguarded, or eat the young eider-ducks and willow grouse for which it seems to have a special fondness; though nothing either young and tender, old and tough, fresh or carrion, goes amiss of its rapacious maw. It is a sea scavenger of more than ordinary capacity, and when faithfully playing in this rôle it lays us under obligation to speak well of it. Certainly the gulls and other sea fowl that eat refuse contribute much to the healthfulness of our coasts.

Before the onslaughts of this black-backed freebooter almost all the tribe of sea fowl quail; and yet, like every other tyrant, it is itself most cowardly, for it will desert even its own young rather than be approached by man, who visits the sins of the father upon the children by pickling them for food when they are not taken in the egg for boiling.

Usually the nest is built with hundreds or even thousands of others on some inaccessible cliff overhanging the sea; or it may be on an island, or on the dunes near the beach, in which latter case it is the merest depression in the turf, lined with grass and seaweed. Two or three—usually three—clay-colored or buff eggs, rather evenly and boldly spotted with chocolate brown, make a clutch. After the nesting season these gulls migrate farther southward than the glaucous gulls, not because they are incapable of withstanding the most intense cold, but because the fish supply is of course greater in the open waters of our coast. With majestic grace they skim along the waves, revealing the dark slate-colored mantle covering their backs like a pall, for which they must bear the gruesome name of "Coffin Carrier."

American Herring Gull
(Larus argentatus smithsonianus)

Called also: WINTER GULL

Length—24 to 25 inches.

Male and Female—In summer: Mantle over back and wings deep pearl gray, also known as "gull blue;" head, tail, and under parts white. Outer feathers of wings chiefly black, with rounded white spots near the tips. Bill bright yellow. Feet and legs flesh-colored. *In winter:* Similar to summer plumage, but with grayish streaks or blotches about the head and neck. Bill less bright.

Young—Upper parts dull ashy brown ; head and neck marked with buff, and back and wings margined and marked with the same color ; outer feathers of wings brownish black, lacking round white spots ; black or brownish tail feathers gradually fade to white.

Range—Nests from Minnesota and New England northward, especially about the St. Lawrence, Nova Scotia, Newfoundland, and Labrador. Winters from Bay of Fundy to West Indies and Lower California.

Season—Winter resident. Common from November until March.

As the English sparrow is to the land birds, so is the herring gull to the sea fowl—overwhelmingly predominant during the winter in the Great Lakes and larger waterways of the interior, just as it is about the docks of our harbors, along our coasts, and very far out at sea; for trustworthy captains declare the same birds follow their ships from port to port across the ocean.

Occasionally at low tide one may meet with a few herring gulls on the sand flats of the beach, feeding on the smaller shell fish half buried there. It is Audubon, the unimpeachable, who relates how these birds, that he so carefully studied in Labrador one summer, break open the shells to extract the mollusks, by carrying them up in the air, then dropping them on the rocks. "We saw one that had met with a very hard mussel," he writes, "take it up three times in succession before it succeeded in breaking it; and I was much pleased to see the bird let it fall each succeeding time from a greater height than before."

Again, one may see a flock of herring gulls "bedded" on the water floating about to rest. All manner of boats pass close beside such a tired company in New York harbor without disturbing it; for these gulls, unlike the glaucous and black-backed species, show little fear of man or his inventions.

But it is high in air, sailing on motionless wings in the wake of an ocean steamer, that one mentally pictures the herring gull. Apparently the loose flock, floating idly about, have no thought beyond the pure sport. Suddenly one bird drops like a shot to the water's surface, spatters about with much wing-flapping and struggle of feet, then, rising again with a small fish or morsel of refuse in its grasp, leads off from a greedy horde of envious companions in hot pursuit that likely as not will overhaul him and rob him of his dinner. Dining abundantly and often, rather than flying about for idle pleasure, is the gull's real business of life.

With all their exquisite poetry of motion, it must be owned that these birds have also numerous prosaic qualities, exercised in their capacity of scavengers. Rapacious feeders, tyrannical to smaller birds that they can rob of their prey, and possessed of insatiable appetites for any food, whether fresh or putrid, that comes in their reach, the gulls alternately fascinate by their grace and animation in the marine picture, and repel by the coarseness of their instincts. However, it is churlish to find fault with the scavengers that help so largely in keeping our beaches free from putrifying rubbish. Doubtless the birds themselves, as their name implies, would prefer herrings were they always available.

Unlike the other gulls, this one, where it has been persistently robbed, sometimes nests in trees, and, adapting its architecture to the exigencies of the situation, constructs a compactly built and bulky home, often fifty feet from the ground, and preferably in a fir or other evergreen. Ordinarily a coarse, loose mat of moss, grasses, and seaweed is laid directly on the ground or on a rocky cliff near the sea. Two or three grayish olive

AMERICAN HERRING GULL.

brown, sometimes whitish, eggs, spotted, blotched, and scrawled with brown, are laid in June. In the nesting grounds the herring gulls are shy of men and fierce in defending their mates and young, to whom they are especially devoted. *Akak, kakak* they scream or bark at the intruder, making a din that is fairly deafening.

Before the summer is ended the baby gulls will have learned to breast a gale, sleep with head tucked under wing when rocked on the cradle of the deep, and follow a ship for the refuse thrown overboard, like any veteran. They are the grayish brown birds which one can readily pick out in a flock of adults when they migrate to our coasts in winter.

Ring-billed Gull
(*Larus delawarensis*)

Length—18.50 to 19.75 inches.

Male and Female—Mantle over back and wings light pearl color, rest of plumage white except in winter, when the head and nape are spotted, not streaked, with grayish brown. Wings have "first primary black, with a white spot near the tip, the base of the inner half of the inner web pearl gray; on the third to sixth primaries the black decreases rapidly and each one is tipped with white." (Chapman.) Bill light greenish yellow, chrome at the tip, and encircled with a broad band of black. Legs and feet dusky bluish green. Immature birds are mottled white and dusky, the dark tint varied with pale buff prevailing on the upper parts, the white below. Tail is dusky, tipped with white and pale gray at the base.

Range—Distributed over North America, nests from Great Lakes and New England northward, especially in the St. Lawrence region, the Bay of Fundy, and Newfoundland; more common in the interior than on the seacoast; winters south of New England to Cuba and Central America.

Season—Common winter visitor.

"On the whole the commonest species, both coastwise and in the interior," says Dr. Elliott Coues. Certainly around the salt lakes of the plains and in limited areas elsewhere in the west it is most abundant, and at many points along the Atlantic coast; but off the shores of the Middle and the Southern, if not also of the New England States, it is the herring gull that seems to predominate, except here and there, as at Washington, for example, where the ring-billed species is locally very common indeed. From Illinois to the Mexican Gulf is also a favorite winter resort.

It is not an easy matter to tell one of these two commonest species from the other, unless they are seen together, when the larger size of the herring gull and the black band around the bill of the ring-billed gull are at once apparent. These birds fraternize as readily as they bully and rob their smaller relations or each other when hunger makes them desperate. One rarely sees a gull alone: usually a loose flock soars and floats high in the air, apparently idle, but in reality keeping their marvelously sharp eyes on the constant lookout for a morsel of food in the waters below. In the nesting grounds countless numbers occupy the same cliffs, and large companies keep well together during the migrations.

Inasmuch as most of the characteristics of the ring-billed gull belong also to the herring gull, the reader is referred to the longer account of the latter bird to save repetition. When living inland the ring-billed gull, beside eating everything that its larger kin devour with such rapacity, catches insects both on the ground and on the wing. A trick at which it is past-master is to follow a school of fish up the river, then, when a fish leaps from the water after a passing insect, swoop down like a flash and bear away fish, bug, and all.

Laughing Gull

(Larus atricilla)

Called also: BLACK-HEADED GULL; RISIBLE GULL

Length—16 to 17 inches.

Male and Female—In summer: Head covered with a dark slate brown, almost black, hood, extending farther on throat than on nape, which is pure white like the breast, tail, and under parts. Mantle over back and wings dark, pearl gray. Wings have long feathers, black, the inner primaries with small white tips. Bill dark reddish, brighter at the end. Eyelids red on edge. Legs and feet dusky red. Breast sometimes suffused with delicate blush pink. *In winter:* Similar to summer plumage, except that the head has lost its hood, being white mixed with blackish. Under parts white without a tinge of rose. Bill and feet duller.

Young—Light ashy brown feathers, margined with whitish on the upper parts, forehead and under parts white, sometimes clouded with dark gray; tail dark pearl gray with broad band of blackish brown across end; primaries black.

Range—"Atlantic and Gulf coasts of the United States, north to Maine and Nova Scotia; south in winter through West Indies, Mexico (both coasts), Central America, and northern South America (Atlantic side), to the Lower Amazon." A. O. U.

Season—Summer resident, and visitor throughout the year.

No bird that must lift up its voice to drown the howlings of the gale and the pounding, dashing surf in an ocean storm might be expected to have a soft, musical call; and the gulls, that pass the greater part of their lives at sea, must therefore depend upon squalls, screams, barks, and shrill, high notes that carry long distances, to report news back and forth to members of the loose flocks that hunt together above the crest of the waves. The laughing gull, however, utters a coarse scream in a clear, high tone, like the syllables *oh-hah-hah-ah-ah-hah-hah-h-a-a-a-a-ah*, long drawn out toward the end and particularly at the last measure, that differs from every other bird note, "sounding like the odd and excited laughter of an Indian squaw," says Langille, "and giving marked propriety to the name of the bird." All gulls chatter among themselves, the noise rising sometimes to a deafening clamor when they are disturbed in their nesting grounds; but the laughing gull, in addition to its long-drawn, clear note on a high key, "sounding not unlike the more excited call-note of the domestic goose," suddenly bursts out, to the ears of superstitious sailors, into the laugh that seems malign and uncanny.

A more southern species than any commonly seen off our shores, the laughing gull nests from Texas and Florida to Maine, though it is not a bird of the interior, as the ring-billed species is, nor so pelagic as the herring gull. It delights in reedy, bush-grown salt marshes that yield a rich menu of small mollusks, spawn of the king crab and other crustaceans, insects, worms, and refuse cast up by the tide. In such a place it also nests in large colonies, forming with its body a slight depression in the sand that is scantily lined with grasses and weeds from the beach, and concealed by a tussock of grasses. Three to five eggs, varying from olive to greenish gray or dull white, profusely marked with chocolate brown, are not so rare a find for the collector as the eggs of most other gulls that nest in the extreme north, where only the hardy explorers in search of the North Pole count themselves more fortunate sometimes to find a square meal of gulls' eggs.

Formerly these laughing gulls were exceedingly abundant all along our coasts. Nantucket was a favorite nesting resort, so were the marshes of Long Island and New Jersey; but unhappily a fashion for wearing gulls' wings in women's hats arose, and though *only* the wings were used, as one woman naïvely protested when charged with complicity in their slaughter, the birds have been all but exterminated at the north. In southern waters they are, happily, common still, and will be again at the north when the beneficent bird laws shall have had time to operate.

Bonaparte's Gull
(Larus philadelphia)

Called also: ROSY GULL

Length—14 inches.

Male and Female—In summer: Head and throat deep sooty slate, the hood not extending over nape or sides of neck, which are white like the under parts and tail. Mantle over back and wings pearl gray. Wings white and pearl gray. Primaries of wings marked with black and white. Bill black. Legs and feet coral red. In nesting plumage only, the white under parts are suffused with rosy pink. *In winter:* Similar, except that the birds lack the dark hood, only the back and sides of the head washed with grayish; white on top.

Young—Grayish washings on top of head, nape, and ears; mantle over back and wings varying from brownish gray to pearl gray; upper half of wings grayish brown; secondaries pearly gray; primaries, or longest feathers, at the end much marked with black; white tail has black band a short distance from end, leaving a white edge showing. Underneath, white.

Range—From the Gulf of Mexico to Manitoba and beyond in the interior; Atlantic and Pacific coasts. Nests north of United States.

Season—Common spring and autumn migrant. A few winter north.

This exquisite little gull, whose darting, skimming flight suggests that of the sea swallow, flies swallow-fashion over the ploughed fields of the interior to gather larvæ and insects, as well as over the ocean to pick up bits of animal food, either fresh or putrid, that float within range of its keen, nervous glance. Jerking its head now this way and now that, suddenly it turns in its graceful flight to swoop backward upon some particle passed a second before. Nothing it craves for food seems to escape either the eyes or the bill of this tireless little scavenger. In sudden freaks of flight, in agility and lightness of motion, it is conspicuous in a family noted for grace on the wing.

A front view of Bonaparte's gull, as it approaches with its long pointed wings outspread, would give one the impression that it is a black-headed white bird, until, darting suddenly, its pearly mantle is revealed. It is peculiarly dainty whichever way you look at it.

In the author's note book are constant memoranda of seeing these little gulls hunting in couples through the surf on the Florida coast one March. Mr. Bradford Torrey records the same observation, but adds, "that may have been nothing more than a coincidence." Is it not probable that these gulls, like all their kin, in their devotion to their mates, were already paired and migrating toward their nesting grounds far to the north? While the birds hunted along the Florida shore they kept up a plaintive, shrill, but rather feeble cry, that was almost a whistle, to each other; and if one was delayed a moment by dipping into the trough of the wave for some floating morsel, it would nervously hurry after its mate as if unwilling to lose a second of its company. In the autumn migrations, however, these "surf gulls," as Mr. Torrey calls them, are seen in large flocks along our coasts, and inland, too, where there is no surf for a thousand miles.

The nest, which is built north of the United States, is placed sometimes in trees, sometimes in stumps, or in bushes, the rude cradle of sticks, lined with grasses, containing three or four grayish olive eggs, spotted with brown, chiefly at the larger end. Such a clutch is a rare find for the collector, few scientists, even, having seen the Bonaparte gulls at home. Charles Bonaparte, Prince of Canino, might have left us a complete life history of his namesake, had not European politics cut short his happy and profitable visit in America.

TERNS, OR SEA SWALLOWS
(Subfamily Sterninæ)

Marsh Tern
(Gelochelidon nilotica)

Called also : GULL-BILLED TERN, OR SEA SWALLOW

Length—13 to 15 inches.

Male and Female—Top and back of head glossy, greenish black; neck all around, and under parts, white; mantle over back and wings, pearl gray; bill and feet black, the former rather short and stout for this family; wings exceedingly long and sharp, each primary surpassing the next fully an inch in length. Tail white, grayish in the centre, and only slightly forked. In winter plumage similar to the above, except that the top of head is white, only a blackish space in front of eyes; grayish about the ears.

Range—"Nearly cosmopolitan; in North America chiefly along the Atlantic and Gulf coasts of the United States, breeding north to southern New Jersey, and wandering casually to Long Island and Massachusetts; in winter both coasts of Mexico and Central America, and south to Brazil." A. O. U.

Season—Summer visitor. Summer resident south of Delaware.

A very common species, indeed, off the coasts of our southern States, this tern, which one can distinguish from its relatives by its heavy black bill and harsh voice, appears at least as far north as Long Island every summer, and occasionally a straggler reaches Maine. While allied very closely to the gulls, that come out of the far north in the winter to visit us, the terns reverse the order and come out of the south in summer.

All manner of beautiful curves and evolutions, sudden darts and dives distinguish the flight of terns, which in grace and airiness of motion no bird can surpass ; but this gull-billed tern is particularly alert and swallow-like, owing to its fondness for insects which must be pursued and caught in mid-air. Fish it by no means despises, only it depends almost never for food upon diving through the water to capture them, as others of its kin do, and almost entirely upon aërial plunges after insects. For this reason it haunts marsh lands and darts and skims above the tall reeds and sedges, also the home of winged bettles, moths, spiders, and aquatic insects, dividing its time between the waving plants and the water waves that comb the beach. It is never found far out at sea, as the gulls are, though rarely far from it.

Like the black tern, it is not a beach-nester, but resorts in companies to its hunting grounds in the marshes, and breaks down some of the reeds and grasses to form what by courtesy only could be termed a nest. Three to five buffy white eggs, marked with umber brown and blackish, especially around the larger end, are usual; but all terns' eggs are exceedingly variable. Once *Anglica* was the specific name of the gull-billed tern; but because our English cousins liked the eggs for food, and used the wings for millinery purposes, the bird is now deplorably rare in England.

"It utters a variety of notes," says Mr. Chamberlain, "the most common being represented by the syllables *kay-wek, kay-wek*. One note is described as a laugh, and is said to sound like *hay, hay, hay*."

Royal Tern
(Sterna maxima)

Called also: CAYENNE TERN; GANNET-STRIKER

Length—18 to 20 inches.

Male and Female—Top and back of head glossy, greenish black, the feathers lengthened into a crest; mantle over back and wings light pearl color; back of neck, tail, and under parts white; inner part of long wing feathers (except at tip) white; outer part of primaries and tip, slate color. Feet black. Bill, which is long and pointed, is coral or orange red. Tail long and forked. After the nesting season and in winter, the top of head is simply streaked with black and white, and the bill grows paler.

Range—Warmer parts of North America on east and west coasts, rarely so far north as New England and the Great Lakes.

Season—Summer visitor. Resident in Virginia, and southward.

Terns

It is the larger Caspian tern, measuring from twenty to twenty-three inches, and not the royal tern, that deserves to be called *maxima*, however imposing the size of the latter bird may be, thanks to its elongated tail; but unless these two birds may be compared side by side in life—a dim possibility—it is quite hopeless for the novice to try to tell which tern is before him.

Off the Gulf shore, especially in Texas, Louisiana, and Florida, where great numbers live, this handsome bird exercises its royal prerogative by robbing the fish out of the pouch of the pelican, that is no match, in its slow flight, for this dashing monarch of the air. But if sometimes tyrannical, or perhaps only mischievous, it is also an industrious hunter; and with its sharp eyes fastened on the water, and its bill pointed downward, mosquito fashion, it skims along above the waves, making sudden evolutions upward, then even more sudden, reckless dashes directly downward, and under the water, to clutch its finny prey. With much flapping of its long, pointed wings as it reappears in an instant above the surface, it mounts with labored effort into the air again, and is off on its eager, buoyant flight. There is great joyousness about the terns a-wing; dashing, rollicking, aërial sprites they are, that the Florida tourists may sometimes see tossing a fish into the air just for the fun of catching it again, or dropping it for another member of the happy company to catch and toss again in genuine play. It would even seem that they must have a sense of humor, a very late appearing gift in the evolution of every race, scientists teach; and so this lower form of birds certainly cannot possess it, however much they may appear to.

While the terns take life easily at all times, nursery duties rest with special lightness. The royal species makes no attempt to form a nest, but drops from one to four rather small, grayish white eggs marked with chocolate, directly on the sand of the beach, or at the edge or a marshy lagoon. As the sun's rays furnish most of the heat necessary for incubation, the mother bird confines her sitting chiefly to her natural bedtime.

Common Tern

(Sterna hirundo)

Called also: WILSON'S TERN; SEA SWALLOW; SUMMER GULL; MACKEREL GULL

Length—14 to 15 inches.

Male and Female—In summer: Whole top of head velvety black, tinged with greenish and extending to the lower level of the eyes and onto the nape of neck. Mantle over back and wings pearl gray. Throat white, but breast and underneath a lighter shade of gray, the characteristic that chiefly distinguishes it from Forster's tern, which is pure white on its under parts. Inner border of inner web of outer primaries white, except at the tip. Tail white, the outer webs of the outer feathers pearl gray. Tail forked and moderately elongated, but the folded wings reach one or two inches beyond it. Legs and feet orange red. Bill, which is as long as head, is bright coral about two-thirds of its length, a black space separating it from the extreme tip, which is yellow. *In winter:* Similar to summer plumage, except that the front part of head and under parts are pure white; also that the bill becomes mostly black. Young birds similar to adults in winter, but with brownish wash or mottles on the back, with slaty shoulders and shorter tail.

Range—"In North America, chiefly east of the plains, breeding from the Arctic coast, somewhat irregularly, to Florida, Texas, and Arizona, and wintering northward to Virginia; also coast of Lower California." A. O. U.

Season—Summer resident. May to October.

Ironically must this particularly beautiful, graceful sea swallow now be called the common tern, for common it scarcely has been, except in the dry-goods stores, since its sharply pointed wings, and often its entire body also, were thought by the milliners to give style to women's hats. Great boxes full of distorted terns, their bills at impossible angles, their wings and tails bunched together, sicken the bird-lover who strolls through the large city shops on "opening day." Countless thousands of these birds must have been slaughtered to supply the demand of thoughtless women in the last twenty years; and although the egret has had its turn of persecution, and that in an especially cruel way, the fashion for wearing terns, either entire or in sections, continues with a hopeless pertinacity that no other mode of hat trimming seems wholly to divert. Chicken feathers, arranged to imitate them, are necessarily accepted as substitutes more and more, however.

Through the efforts of Mr. Mackay, of Nantucket, the terns are at last protected on a number of low, sandy islands adjacent to his home, where nesting colonies had resorted from the earliest recollection until they were all but exterminated by the companies of men and boys who sailed over from the mainland to collect plumage and the delicately flavored eggs. Muskegat and Penekese Islands, off the extreme southeastern end of Massachusetts—the latter made famous by Agassiz—and Gull Island, off the Long Island coast, the only nesting grounds left these sea swallows in the north, are now guarded by paid keepers, who see to it that no unfriendly visitor sets foot on the shores until the downy chicks are able to fly in September. It was mainly through the efforts of Mr. William Dutcher that the terns were taken under the protection of the A. O. U., the Linnæan Society, and the A. S. P. C. A., at Gull Island. In May the terns begin to arrive from the south, having apparently mated on the journey. Little or no part of the honeymoon is spent in making a nest, as any little accumulation of drift, or the bare sand itself, will answer the purpose of these shiftless merry-makers that no responsibilities can depress nor persecution harden. Lightness and grace of flight, as well as of heart, are their certain characteristics. Before family cares divert them, in June, how particularly lively, dashing, impetuous, exultant, free, and full of spirit they are! A sail across to the terns' nesting grounds is recommended to those summer visitors who sit about on the piazzas complaining of *ennui* at Nantucket, Martha's Vineyard, and Shelter Island.

As a boat approaches a nesting colony on one of the few low, sandy islands where one may be still found, a canopy or cloud of birds spreads overhead—a surging mass of excited creatures, darting, diving in a maze without plan or direction, like a flurry of huge snowflakes through the summer sky. The air fairly vibrates with the sharp, rasping notes of alarm uttered in a mighty chorus of complaint, very different from the almost musical call, half melancholy, half piping, that the birds continually utter when undisturbed. If the visit be made to the island in June, the upper beach, above the reach of tide, will be

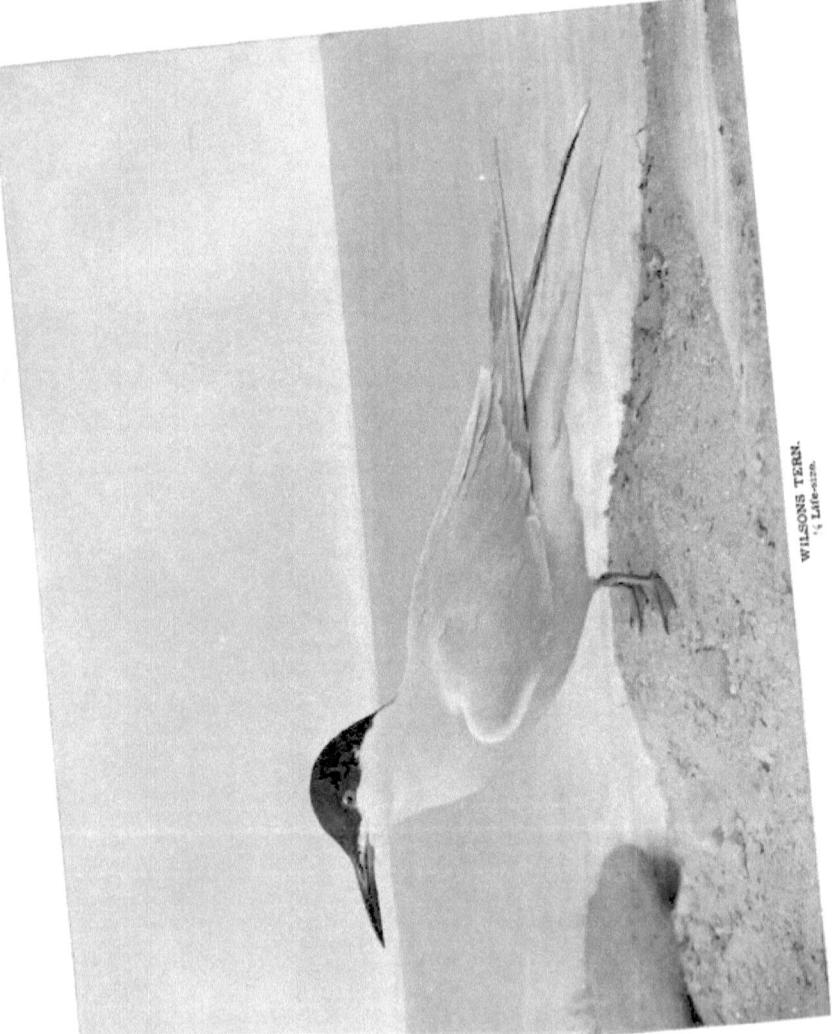

WILSONS TERN.
½ Life-size.

scattered over here and there with clutches of eggs that so closely imitate the speckled sand, one is apt to step on them unawares. Only the slightest depression, lined with a wisp of grass or bit of seaweed, is made in pretense of a nest; and as the gay mothers leave the work of incubating chiefly to the sun, confining themselves only at night or during storms, the visitor may be forgiven if the sound of a crushed shell under foot is his first intimation of a nest among the dried seaweed or beach grass among the rocks. It was Audubon who said there were never more than three eggs in a nest; but Mr. Parkhurst, at least, has found four.

Should the visitor reach the island in July, he will find great numbers of downy young chicks running about, but quite dependent on their parents for grasshoppers, beetles, small fish, and smaller insects that are the approved diet for young terns. The young are tame as chickens; but the old birds at this time are especially bold and resentful of intrusion. Darting down to a clamoring chick, a parent thrusts a morsel down its throat without alighting, and is off again for more, and still more. Later the food is simply dropped for the fledglings to help themselves. Still later, little broods are led to the ocean's edge, sand shoals, or the marshes, to hunt on their own account; and by September, old and young congregate in great groups to follow the movement of the blue fish, that pursue the very small fish, "shiners," that they also feed on.

But whether flirting, nesting, hunting, or flying at leisure, there is a refreshing joyousness about the tern that makes it a delight to watch. In the very excess of good spirits one will plunge beneath the water after a little fish, then mounting into the air again, it will deliberately drop it from its bill for another tern to dash after, and the new possessor will toss it to still another member of the jolly flock, and so keep up the game until the fish is finally swallowed. It has been suggested that terns go through this performance to kill the fish, as a cat plays with a mouse; but it is only occasionally they play the game of catch and toss, and when all the company seem to be in the mood for the fun.

Another beautiful sight is the pose of a tern just before alighting, when, with long, pointed wings held for a moment high above its back, they flutter like the wings of a butterfly. But then it would be difficult to name a posture of this graceful bird that is not beautiful, unless we except the act of scratching its head with one foot while on the wing; and this is, perhaps, more amusing than lovely. This sea swallow also has the accomplishment of opening and shutting its tail like a fan, so that one moment it will look like a single pointed feather, and the next it may be narrowly forked or widely stretched into an open triangle. While flying, the birds are exceedingly watchful, jerking their heads now this way, now that, with nervous quickness, all the time keeping their "bill pointing straight downward, which makes them look curiously like colossal mosquitoes," to quote Dr. Coues's famous comparison. By the middle of October the terns migrate southward from the New England and Long Island waters to enjoy the perpetual summer, of which they seem to be a natural exponent.

Roseate Tern
(Sterna dougalli)

Called also: PARADISE TERN

Length—14.50 to 15.50 inches.

Male and Female—In summer: Mantle over back and wings delicate pearl color, lighter and fading to white on the tail, which is exceedingly long and deeply forked. Feathers on crown, which reaches to the eyes and the back of neck, are black and long. Under parts white, tinted with rose color. Long, slender black bill, reddish at the base and yellow at the tip. Feet and legs yellowish red. *In winter:* Under parts pure white, having lost the rose tint; forehead and cheeks white. Crown becomes brownish black, mixed with white; some brownish feathers on wings; pearl gray tail, without extreme elongation or forking.

Range—Temperate and warm parts of Atlantic coast, nesting as far north as New England; most abundant, however, south of New Jersey. Winters south of United States.

Season—Comparatively rare summer resident at the north, but regular.

Closely associated with the common tern in their nesting colonies on Gull and Muskegat Islands, described in the preceding biography, this most exquisite member of all the family may be distinguished from its companions by the very long and sharply pointed tail feathers, and the lovely rose-colored flush it wears on its breast as a sort of wedding garment. This tint is all too transitory, however; family cares fade it to white; death utterly destroys it, though it sometimes changes to a salmon shade as the lifeless body cools, before disappearing forever. Comparatively short of wing, the roseate tern cannot be said to lose any of the buoyancy and grace of flight, the dash and ecstasy that give to the movements of all the tribe their peculiar fascination.

It has been said that these birds' eggs are paler than those of the common terns, which are very variable, ranging from olive gray or olive brownish gray to (more rarely) whitish or buff, heavily marked with chocolate; but though they may average paler, many are identical with those just described; and as the birds nest in precisely the same manner, on the same beach, not even an expert could correctly name the egg every time without seeing the adult bird that laid it identify its own.

A single harsh note, *cack*, rises above the din made by the common terns, and at once identifies the voice of the roseate species. It would be unfair to attribute the melancholy, unpleasing quality of the terns' voices to their dispositions, which we have every reason to suppose are particularly joyous and amiable. This bird also appears less excitable; but in all other particulars than those already noted the common and the roseate terns share the characteristics described in the preceding account, to which the reader is referred. It is a gratification to know that at the close of the first season, when the tern colony had been protected at Gull Island, Mr. Dutcher could report an increase of from one thousand to fifteen hundred birds, virtually an increase of one half the total number in one year.

With the four species of tern that nest in the neighborhood of New York and New England, the Arctic Tern (*Sterna paradisæa*) has nearly all characteristics in common, and the few peculiarities that differentiate it from the common tern are quickly learned. While these birds are similar in color, the Arctic tern "differs in having less gray on the shaft part of the inner web of the outer primaries, in having the tail somewhat longer, the tarsi and bill shorter; while the latter, in the adult, is generally without a black tip." (Chapman.) Its voice is shriller, with a rising inflection at the end, and resembling the squeal of a pig; but it also has a short, harsh note that can scarcely be distinguished from the roseate tern's cry.

In habits the Arctic tern is said to have the doubtful peculiarity of being more bold in defense of its young than any of its kin; first in war, most fierce in attack, and the last to leave an intruder. At Muskegat Island, where great colonies of terns regularly nest and are protected under the wing of the law (see page 50) it is usually the Arctic tern that dashes frantically downward into the very face of the visitor who dares to inspect its eggs. These are of a darker ground and more heavily marked than those of the common tern. Mr. Chamberlain says these terns "may be seen sitting on a rock or stump, watching for their prey in kingfisher fashion. They float buoyantly on the surface, but rarely dive beneath the water." Their nesting range is from Massachusetts to the Arctic regions; and they winter southward only to Virginia and California.

Least Tern

(Sterna antillarum)

Called also: SILVERY TERN; LITTLE STRIKER

Length—9 inches.

Male and Female—*In summer:* Glossy greenish black cap on head, with narrow white crescent on forehead, and extending over the eyes. Cheeks black. Mantle over back, wings, and tail, pearl gray. A few outer wing feathers, black. Under parts satiny white. Bill, about as long as head, is yellow, tipped with black. Feet and legs, orange. Tail moderately forked. *In winter:* Top of head white, with black shaft lines on feathers. Mantle darker than in summer; a band of grayish black along upper wing, and most of the primaries black. Feet paler; bill black.

Range—Northern parts of South America, up the Pacific coast to California, and the Atlantic to Labrador; also on the larger bodies of water inland. Nests locally throughout its range. Winters south of United States.

Season—Irregular migrant and summer visitor.

Any of the thirteen species of terns that we may call ours is easily the superior of this little bird in size; but in grace and buoyancy of flight, in dash and impetuosity, it certainly owns no master among its own accomplished kin, and suggests the movements of the swallow alone among the land birds. Skimming just above the marshes near the sea or inland waters, as any swallow might, to feed upon the dragon-flies and other winged insects that dart in and out of the sedges, this little tern flashes its silvery breast in the sunlight, swallow fashion, and appears to have the "sandals of lightning on its feet" and "soft wings swift as thought" sung of by Shelley.

Off the shores of the low, sandy islands on the extreme southeastern coast of Massachusetts, where these terns nest regularly, though in sadly decreased numbers, they may be seen in company with the common tern, the roseate and the Arctic species, that also make their summer home there, as the joyous birds hunt in loose flocks together above the waves. There can be no difficulty in picking out the dainty, elegant little figure that floats and skims in mid-air, with bill pointing downward as if it were a lance to spear some tiny fish swimming in the ocean below.

Hovering for an instant on widely outstretched wings, like a miniature hawk, the next instant it has suddenly plunged after its prey, to reappear with it in its bill, since its feet are too webbed and weak to carry anything; and, if the season be midsummer, it will doubtless head straight for its nest on the sand, to drop its spoils in the midst of a brood of three or four very tame young fledglings. In Minnesota, Dakota, and other inland states, both old and young birds feed almost entirely on insects.

All terns keep so closely within the lines of family traditions that a description of one member answers for each, with a few minor changes; and the reader is referred to the life history of the common tern for fuller particulars of the least species, to avoid constant repetition. Although this little bird nests directly on the sand, leaving the greater part of its incubating duties to the sun, as other terns do, its eggs may be easily distinguished, which is not true of the others, because of their smaller size and buffy white, brittle shells that are often wreathed with chocolate markings around the larger end, the rest of the egg being plain.

Some one has described the bird's voice as "a sharp squeak, much like the cry of a very young pig."

Black Tern

(Hydrochelidon nigra surinamensis)

Called also: SHORT-TAILED TERN

Length—9.50 to 10 inches.

Male and Female—In summer: Head, neck all around, and under parts jet black, except the under tail coverts, which are white. Back, wings, and tail slate color. *In winter:* Very different, forehead, sides of head, nape, and under parts white; under wing coverts only, ashy gray; back of the head mixed black and white; mantle over back, wings, and tail, deep pearl gray. Many feathers with white edges. In the process of molt, head and under parts show black and white patches. Immature specimens resemble the winter birds, except that their upper parts are more or less mixed with brownish, and their sides washed with grayish.

Range—North America at large, in the interior and along the coasts, but most abundant inland; nests from Kansas and Illinois northward, but not on the Atlantic coast.

Season—Irregular migrant on the Atlantic coast from Prince Edward's Island southward. Common summer resident inland. May to August or September.

Although eastern people rarely see this dusky member of a tribe they are wont to think of as having particularly delicate pearl and white plumage, it is the most abundant species in the west, and indeed the only one of the entire order of long-winged swimmers that commonly nests far away from the sea in the United States. Early in May it arrives in large flocks that have gathered on the way from Brazil and Chile to nest in the Middle States, west of the Alleghanies, and northward. A large colony takes up its residence in the fresh-water marshes and reedy sloughs so abundant in southern Illinois and elsewhere in the middle west; and although the birds have apparently mated during the migration, if not before, there are many flirtations and petty jealousies exhibited before family cares banish all nonsense in June. Not that the bird makes any effort to construct a nest, in which case it could hardly be a tern at all, so easy-going are all the family in this respect; nor that it is depressed by long, patient sittings on the eggs, for the incubating is, for the most part, left to the sun, when it shines, but all terns are devoted parents, however emancipated they are from much of the parental drudgery. Sometimes the eggs are laid directly on the wet, boggy ground; others in a saucer-shaped structure of decayed reeds and other vegetation, often wet and floating about in the slough; and again they have been found in better constructed, more compact cradles, resting on the flat foundation of the home of the water rat. The eggs are two or three, grayish olive brown, sometimes very pale and clean, marked with spots and splashes of many sizes, but chiefly large and bold masses that have a tendency to encircle the larger end.

To visit a marsh when several hundred of these aquatic nests keep the cloud of dusky little parents in a state of panic, is to become deaf and dazed by the terrific din of harsh, screaming cries uttered by the little black birds that encircle one's head, menacing, darting, yet doing nothing worse than needlessly tormenting themselves. Retreat to a good point of vantage to watch the colony, and it quickly regains its lost confidence to the point of ignoring your presence; and the jolly company skim, soar, hover on outstretched wings, then dart in and out in a pathless maze that fascinates the sight. The flight is exquisite, swift, graceful, buoyant, and apparently without the slightest effort. Occasionally a bird will descend from the aërial game, and, checking its flight above its nest, poise for an instant on quivering wings, held high above its back, as if it spurned the earth.

Doubtless the diet of insects, which must be pursued and captured on the wing in many cases, cultivates much of the dash and impetuosity so characteristic of this tern. Fish appear to form no part of its bill of fare. It may "frequently be seen dashing about in a zig-zag manner," writes Thompson in his "Birds of Manitoba," and "so swiftly the eye can offer no explanation of its motive until . . . a large dragon-fly is seen hanging from its bill." Beetles, grasshoppers, and aquatic insects of many kinds encourage other extraordinary feats of flight. Mr. Thompson tells of meeting these birds far out on the dry, open plains, scouring the country for food at a distance of miles from its nesting ground. John Burroughs once had brought to him, to identify, a sooty tern, a near relative of the black species, that a farmer had picked up exhausted and emaciated in his meadow, fully one hundred and fifty miles from the sea, and at least two thousand miles from the Florida Keys, the bird's chosen habitat.

Terns

It had starved to death, he says, "ruined by too much wing. Another Icarus. Its great power of flight had made it bold and venturesome, and had carried it so far out of its range that it starved before it could return."

By the end of July the young black terns have sufficiently developed to join the flocks of adults that even thus early show the restlessness called forth by the instinct for migration. In August migration commences in earnest; and when we see the birds east of the Alleghanies, they are usually on their journey south, the only time they show a preference for the Atlantic coast.

BLACK TERN.
Mother and Young with Egg.

SKIMMERS
(Family Rynchopidæ)

Black Skimmer
(Rynchops nigra)

Called also: SCISSOR BILL; CUT-WATER

Length—16 to 20 inches.

Male and Female—Crown of head, back of neck, and all upper parts, glossy black; forehead, sides of head and neck, and under parts white, the latter suffused with cream or pale rose in the nuptial season. Lining of wings black. Broad patch on wing, the tips of the secondaries, white; also the outer tail feathers, while the inner ones are brownish. Lower half of bill, measuring from 3.50 to 4.50 inches, is about one inch larger than upper half. Basal half of bill carmine; the rest black. Bill rounded at the ends, and compressed like the blade of a knife. Feet carmine, with black claws.

Range—"Warmer parts of America, north on the Atlantic coast to New Jersey, and casually to the Bay of Fundy." A. O. U.

Season—May to September. Summer resident so far north as New Jersey; a transient summer visitor beyond.

Closely related as the skimmers are to both gulls and terns, it is small **wonder the** three species constituting this distinct family should be honored by a separate classification on account of the extraordinary bill that is their chief characteristic. "Among the singular bills of birds that frequently excite our wonder," says Dr. Coues, "that of the skimmers is **one** of the most anomalous. The under mandible is much longer than the upper, compressed like a knife-blade; its end is obtuse; its sides come abruptly together and are completely soldered; the upper edge is as sharp as the under, and fits a groove in the upper mandible; the jawbone, viewed apart, looks like a short-handled pitchfork. The upper mandible is also compressed, but less so, nor is it so obtuse at the end; its substance is nearly hollow . . . and it is freely movable by means of an elastic hinge at the forehead."

But curious as the bill is when one examines a museum specimen, it becomes vastly more interesting to watch in active use on the Atlantic. The black skimmer, the only one that visits our continent, happily keeps close enough to shore when hunting for the small fish, shrimps, and mollusks that high tide brings near, for us to observe its operations. With leisurely, graceful flight, though with frequent flapping of its very long wings, the bird floats and balances just over the water, and as it progresses over a promising shoal teeming with living food, suddenly the lower half of the bladelike bill drops down just below the surface of the water, and with increased velocity of flight the bird literally "plows the main," as Mr. Chapman has said, and receives a rich harvest through the gaping entrance. Thus cutting under or grazing the surface, with the fore part of its body inclined downward, the skimmer follows the plow into the likeliest feeding grounds, which are the estuaries of rivers, sandy shoals, inlets of creeks, the salt marshes, and around the floating "drift" of the beaches. Though strictly maritime, it never ventures out on mid-ocean like the gulls and petrels. From Atlantic City, Cape May, and southward to Florida, the skimmer is an uncommon though likely enough sight to cause a genuine sensation when discovered at work. It is also credited with using its bill as a sort of oyster knife to open mollusks.

Flocks of skimmers come out of the tropics in May, and, like the terns, choose a sandy shore for their nesting colony, and, like the terns again, construct no proper nest for the three or four buffy white, chocolate-marked eggs that are dropped on the sand, high up on the beach, among the drift and shells. Incubating duties rest lightly with the skimmers, also, while the sun shines with generating warmth, so that the natural bedtime of the mother is all the confinement she endures unless the weather be stormy. In September the young birds are able to migrate long distances, although for several weeks after they are hatched they must be fed and tended by their parents; the only use they have for their wings during June and July, apparently, being to stretch them while basking in the sun on the beach. The voice of the skimmer, like that of the tern, is never so harsh and strident as during the nesting season.

It seems odd that birds so long and strong of wing as these should hug the coast so closely and not venture out on the open seas, until we consider the nature of their food and the probability of starvation in deep waters.

TUBE-NOSED SWIMMERS

Shearwaters
Petrels

TUBE-NOSED SWIMMERS

Shearwaters and Petrels
(Order Tubinares)

The albatrosses, fulmars, shearwaters, and petrels, that comprise this order of water-birds, live far out on the ocean, touching land only to nest, and are unsurpassed in powers of flight, owing to the constant exercise of their long, strong, pointed wings. None of our American sportsmen can wail, with Coleridge's Ancient Mariner, that he "shot the albatross," for the several species that comprise its family *(Diomedeidæ)* confine themselves to the southern hemisphere. The wandering albatross, the largest of all sea birds, with a wing expanse of from twelve to fourteen feet, and "Mother Carey's chickens," the little petrels that travellers on the north Atlantic frequently see, represent the two extremes of size among the pelagic birds.

The plumage of birds of this order is compact and oily, to resist water, and differs neither in the sexes, nor at different seasons, so far as is known. Sooty black, grays, and white predominate. The peculiarity of nostrils, tubular in form, and nearly always horizontal, divide the birds into a distinct order.

Shearwaters and Petrels
(Family Procellariidæ)

"Mother Carey's Chickens" may be distinguished by their small size, slight, elegant form, and graceful, airy, flickering flight, as contrasted with the strong, swift flying of the larger shearwaters that often sail with no visible motion of the pinions. Birds of the open sea, feeding on animal substances, particularly the fatty ones, they may sometimes be noticed in flocks, picking up the refuse thrown overboard from the ship's kitchen, on the ocean highway, like the more common herring gull. They seem

Tube-nosed Swimmers

to be ever on the wing, though their webbed feet indicate that they must be good swimmers when they choose. Hardly any birds are less known than all these ocean roamers and their kin that come to land only to nest. The nest and eggs of the common shearwater, that wanders over the whole Atlantic from Greenland to Cape Horn and the Cape of Good Hope, that sailors often see in flocks of thousands, have yet to be discovered. Petrels burrow holes in the ground like bank swallows.

                    Greater Shearwater.
                    Wilson's Stormy Petrel.
                    Leach's Petrel.

SHEARWATERS AND PETRELS
(Family Procellariidæ)

Greater Shearwater
(Puffinus major)

Called also: HAGDON; WANDERING SHEARWATER; COMMON ATLANTIC SHEARWATER

Length—19 to 20 inches.

Male and Female—Upper parts dark grayish brown. The feathers, except when old, edged with lighter brown; the wings and tail darkest; lightest shade on neck; the white feathers of the fore neck abruptly marked off from the dark feathers of the crown and nape. Under parts white, shaded with brownish gray on sides; under tail coverts ashy gray; upper coverts mostly white. Wings long and pointed. Bill, which is dark horn color, is about as long as head, and has a strong hook at the end. Legs and feet yellowish pink or flesh color.

Range—Over the entire Atlantic Ocean, from Cape Horn and Cape of Good Hope to Arctic Circle.

Season—Irregular visitor to our coast; abundant far off it in winter.

Off the banks of Newfoundland and southward, passengers on the ocean liners sometimes see immense flocks of these birds, smaller than gulls, though larger than pigeons, flying close over the waves, in a direct course, with strong wing beats, then floating often half a mile with no perceptible motion of the wings. The stronger the gale blows, the more does the shearwater seem to revel in it; for as the waves are lifted high enough to curl over in a thin sheet, allowing the light to strike through, the tiny fish are plainly revealed, and quick as thought the bird dives through the combing crest to snap up its prey. Any small particles of animal food cast up by the troubled waters are snatched at with spirit, while with uninterrupted flight the shearwater sweeps

Shearwaters and Petrels

over the waves in wide curves, now deep in the trough, now high above the great swells breaking into foam; but always with "its long, narrow wings set stiffly at right angles with the body," to quote Brewster. Sir T. Browne, who was the first to speak of this bird or its immediate kin, wrote a quaint account of it which is still preserved in the British Museum. "It is a Sea-fowl," he says, "which fishermen observe to resort to their vessels in some numbers, swimming *(sic)* swiftly too and fro, backward, forward and about them, and doth, as it were, *radere aquam*, shear the water, from whence, perhaps, it had its name." No doubt the venerable ornithologist meant to say skimming instead of swimming, for the shearwater almost never rests on the water, except, as is supposed, after dark, to sleep. So characteristic is this constant roving on the wing, that the Turks around the Bosphorus, where these birds have penetrated, think they must be animated by condemned human souls; hence the name *Ames damnées* given the poor innocents by the French. Indeed, all we know about these birds is from hasty glances as they sweep by us at sea; for, although common immediately off our coast in winter, they are never seen to alight on it; and as for either the bird's nest, eggs, and fledglings, they are still absolutely unknown to scientists. A species that is abundant off Australia burrows a hole in the ground near the shore and deposits one pure white egg at the end of the tunnel, just as many petrels do; and it is reasonable to suppose the greater shearwater makes a similar nest. Some white eggs received from Greenland are thought to belong to this species.

Wilson's Stormy Petrel

(Oceanites oceancius)

Called also: MOTHER CAREY'S CHICKEN; DEVIL'S BIRD

Length—7 inches. Very long wings, with an extent of 16 inches, give appearance of greater size.

Male and Female—Upper parts, wings, and tail sooty black; paler underneath, and grayish on wing coverts. The upper tail coverts and frequently the sides of rump and base of tail, white. Bill and feet black. Legs very long, and webs of toes mostly yellow. Tail square and even.

Range—Atlantic Ocean, North and South America, nesting in

WILSONS PETREL.
⅓, Life-size.

southern seas (Kerguelen Island) in February; afterward migrating northward.

Season—Common summer visitor off the coast of the United States.

This is the little petrel most commonly seen off the coast of the United States in summer, silently flitting hither and thither with a company of its fellows like a lot of butterflies in their airy, hovering flight. Owing to the spread of their long wings they appear much larger than they really are, for in actual size the birds are only a trifle longer than the English sparrow, and look like the barn swallow; yet these tiny atoms of the air spend their "life on the ocean wave," and have "their home on the rolling deep,"

> "O'er the deep! o'er the deep!
> Where the whale and the shark and the swordfish sleep—
> Outflying the blast and the driving rain,"

like the stormy petrel of the east Atlantic *(Procellaria pelagica)*, an even smaller species, which doubtless was the bird "Barry Cornwall" had in mind when he wrote his famous verses. Those who go down to the sea in ships are familiar with the petrels that gather in flocks in the wake of the vessel, coursing over the waves, now down in the trough, now up above the crest that threatens to break over their tiny heads; half leaping along a wave, half flying as their distended feet strike the water, and they bound upward again; darting swallow-fashion and skimming along the surface, or flitting like a butterfly above the refuse thrown overboard from the ship's galley. "But the most singular peculiarity of this bird," to quote Wilson, for whom it was named, "is its faculty of standing, and even running, on the surface of the water, which it performs with apparent facility. When any greasy matter is thrown overboard, these birds instantly collect around it, and face to windward, with their long wings expanded, and their webbed feet patting the water, which the lightness of their bodies and the action of the wind on their wings enable them to do with ease. In calm weather they perform the same manœuvre by keeping their wings just so much in action as to prevent their feet from sinking below the surface." It is this appearance of walking on the waves, like the Apostle Peter, that has caused his name to be applied to them.

Particles of animal matter, particularly anything fat or oily, are what the petrels are searching for when they follow a ship; and seeing any such they quickly settle down to enjoy it, then rising again, soon overtake a vessel under steam. Their wing power is marvellous, yet when a gale is blowing in full blast at sea, these little birds are often blown far inland; the capped petrel, for example, that has its proper home in Guadeloupe, in the West Indies, having been found in the interior of New York state after a prolonged "sou'easter." The petrels swim little, if any, though their webbed feet are so admirably adapted for swimming, which might be a greater protection to them than flying when the storms blow. The lighthouses attract many to their death on the stern New England coast.

As night approaches the birds show signs of weariness from the perpetual exercise; for not only have they kept pace with a steamer through the day, but they have made innumerable excursions far from the ship, and played from side to side with a flock of companions at hide-and-go-seek or cross-tag until the eye tires of watching them. But by the time it is dark the last one of the merry little hunters has settled down upon the waves, with head tucked under wing, to rest until dawn while "rocked in the cradle of the deep"; yet it is apparently the very same flock of birds that are busily looking for breakfast the next morning in the wake of the ship, which they must have overtaken with the wings of Mercury.

It would seem these innocent sea-rovers might escape persecution at the hands of man; but an English globe-trotter tells of seeing not only sailors, but passengers, too, who ordinarily feel only *camaraderie* for other fellow travellers on a lonely vessel, shoot these tiny waifs hovering about the ship, to break the tediousness of a long voyage. With the guilty consciences such sailors must have, it is small wonder the petrel is a bird of ill omen to them. They claim it is a harbinger of storms, like its large relative the albatross; and it might easily be, for it delights in rough weather that brings an abundance of food to the surface. All the gruesome superstitions which sailors have clustered around the birds of this entire family, in fact, were woven by Coleridge into his "Rime of the Ancient Mariner."

According to Brünnich, the Faro Islanders draw a wick through the body of the petrel, that is oily from the eating of much fat, and burn the poor thing as a lamp.

Among the many senseless stories sailors tell of the petrel is that it never goes ashore to nest, but carries its solitary egg under its wing until hatched. But the members of the Transit of Venus expedition in the Southern Ocean, several years ago, discovered a large colony of these birds nesting on Kergulen Island. Heretofore, ornithologists, misled by Audubon, had confounded the nest of Wilson's with that of Leach's petrel. Nests containing one white egg each were found in the crevices of rock during January and February. In the latter month the author has seen the birds in great numbers off the Azores, but, unhappily, not on them, for the steamer did not stop there; however, it is not unlikely they nest on these islands, which would seem a convenient rallying place for the birds from the African coast and those that course along the Western Atlantic from Labrador to Patagonia. The young birds are fed by that disgusting process known as regurgitation, that is, raising the food from the stomachs by the parents, which Nuttall says sounds like the cluttering of frogs. Baskett writes in his "Story of the Birds" : "The baby petrel revels in the delights of a cod-liver-oil diet from the start."

Ordinarily quite silent birds, these petrels sometimes call out *weet, weet*, or a low twittering chirp that might be written *pe-up*. But it is near its nest that a bird is most noisy ; and until very recently the home life of this common petrel was absolutely unknown.

. . . .

Leach's, the White-rumped, or the Forked-tailed Petrel, as it is variously known *(Oceandroma leucorhoa)* was the bird carefully studied by Audubon, but confused by him with Wilson's petrel, in which mistake many ornithologists followed him. In size and plumage the birds are almost identical, but the forked tail of Leach's petrel is its distinguishing mark. The outer tail feathers are fully a half inch longer than the middle pair, making the bird look more swallow-like even than Wilson's.

Leach's petrels, while quite as common on the Pacific coast as on the Atlantic, have their chief nesting sites in the Bay of Fundy, while a few nest off the coast of Maine; for it is a more northern species than Wilson's, Virginia and California being its southern boundaries. Nevertheless it is by no means so common off the coast of New England and the Middle States, except around the lighthouses, as Wilson's petrel, that must migrate thousands of miles from the Southern Ocean to pass its summer with us.

Audubon noted that these petrels were seldom seen about their nesting sites during the day, but seemed to have some nocturnal proclivities; for they approached the shore after dark, and flew around like so many bats in the twilight, all the while uttering a wild, plaintive cry. But Chamberlain claims that one of the birds, usually the male, sits on its egg all day while its mate is out foraging at sea. "When handled," he says, "these birds emit from mouth and nostrils a small quantity of oil-like fluid of a reddish color and pungent, musk-like odor. The air at the nesting site is strongly impregnated with this odor, and it guides a searcher to the nest." Sailors have dubbed them with numerous vile names on account of this peculiar means of defense.

A few bits of sticks and grasses laid at the end of a tunnel burrowed in the ground, at the top of an ocean cliff, very much as the bank swallow constructs its nest, make the only home these sea-rovers know. Such a tunnel contains one egg, about an inch to an inch and a half long, and marked, chiefly around the larger end, with small reddish-brown spots. In most respects Leach's petrel is identical with Wilson's, and the reader is therefore referred to the fuller account of that bird.

TOTIPALMATE, OR FULLY WEBBED SWIMMERS

Cormorants

TOTIPALMATE, OR FULLY WEBBED SWIMMERS

(Order Steganapodes)

Birds of this order belong chiefly to tropical or sub-tropical countries, and include the tropic birds, gannets, darters, cormorants, pelicans, and man-o'-war birds, representatives of each of these seven families at least touching our southern coast line, although only the cormorant is common enough north of the southern states to come within the scope of this book. The characteristic that separates these birds into a distinct order is the complete webbing of all the toes; the hallux, or great toe, which in many water-birds is either rudimentary, elevated, or disconnected from the other webbed toes, is in these species flat and fully webbed like the rest, a characteristic no other birds have.

Cormorants

(Family Phalacrocoracidæ)

More than half of all the birds of the order of fully webbed swimmers are cormorants; found in all parts of the world; but of these we have only one, commonly found in the United States around bodies of fresh water inland as well as off the Atlantic coast. Cormorants nest in great colonies and are gregarious at all times. The Chinese have turned their abnormal appetite for fish to good account, by partly domesticating their common species, putting a tight collar around the bird's throat to prevent it from swallowing its prey, and then sending it forth to hunt for its master.

Birds of this family are strong fliers, and although they keep rather close to the water when fishing, often pursuing their game below the surface, they fly high in serried ranks, a few birds deep, but in a long line, during the migrations.

Totipalmate, or Fully Webbed Swimmers

The hooked bill that helps hold a slippery fish secure; the iridescent black and brown plumage, which is the same in both sexes; and certain special featherings of a temporary character that are worn during the nesting season only, are among the most noticeable characteristics of this family.

Double-crested Cormorant.

CORMORANTS
(Family Phalacrocoracidæ)

Double-Crested Cormorant
(Phalacrocorax dilophus)

Called also : SHAG

Length—30 to 32 inches.

Male and Female—Head, neck, lower back, and under parts glossy, iridescent black, with greenish reflections; back and wings light grayish brown, each feather edged with black. A tuft of long, thin black feathers either side of the head, extending from above the eyes to the nape of neck. Birds of the interior show some white feathers among the black ones, while Pacific coast specimens, it is said by Chamberlain, wear wholly white wedding plumes. Wedge-shaped black tail, six inches long, is composed of twelve stiff feathers. Bill longer than head, and hooked at end. Naked space around the eye; base of bill and under throat orange. Legs and feet black; all four toes connected by webs. Winter birds lack the plumes on sides of head, and show more brownish tints in plumage.

Range—North America, nesting from the Great Lakes, Minnesota, Dakota, and Nova Scotia northward; wintering in our southern States south of Illinois and Virginia.

Season—Chiefly a spring and autumn migrant, except where noted above.

Which of the cormorants it was that the Greeks called phalacrocorax, or bald raven, and is responsible for the unpronounceable name borne by the family to this day, is not now certain; but of the thirty species named by scientists, we are at least sure it was not the double-crested cormorant which is peculiar to America. Some of the Latin peoples, thinking the bird suggests by its plumage and its voracious appetite a marine crow (corvus marinus), have given it various titles from which the

Cormorants

English tongue has corrupted first corvorant, then cormorant, whose significance we do not always remember.

Long, serried ranks of double-crested cormorants come flying northward from the Gulf states in April, and pass along the Atlantic shores so high overhead that the amateur observer guesses they are large ducks from their habit of flight, not being able to distinguish their plumage. In the interior of the United States, as well as on the coast, they make frequent breaks in the long migration to their northern nesting grounds, when, if we are fortunate enough, we may watch their interesting hunting habits. Flying low, or just above the surface of the water, the cormorant, suddenly catching sight of a fish, dives straight after it; darts under water like a flash; pursues and captures the victim, though to do it, it must sometimes stay for a long time submerged, then reappears with the fish held tightly in its hooked beak, from which there is no escape. Before the prize is swallowed it is first tossed in the air, then as it descends head downward it lands in the sack or dilatable skin of the cormorant's throat, there to remain in evidence from without until, partly digested, it passes on to the lower part of the bird's stomach. After its voracious appetite has been appeased, the cormorant appears moody and glum.

On the shores of inland waters, particularly, the cormorant often seeks a distended branch of some tree overhanging the lake or river, to sit there, a sombre, meditative figure, only intent on the fish below. In "Paradise Lost," after likening Satan to a wolf preying upon lambs in the sheepfold, Milton continues with another simile :

> " Thence up he flew, and on the tree of life,
> The middle tree, and highest there that grew,
> Sat like a cormorant : yet not true life
> Thereby regained, but sat devising death
> To them who lived."

In Milton's day it was royal sport to go a-fishing with half-domesticated, trained cormorants. A strap was fastened around the bird's throat tight enough to keep it from swallowing its legitimate prey, but loose enough for it to take a full breath. Then it was released to furnish amusement for the royal company assembled on the shore as it darted like an arrow through the clear waters, hunted the fish out of their holes, pursued, captured them, and brought them squirming to its master's feet. A few English noblemen still divert themselves with this mediæval pastime, according to Professor Alfred Newton of Cambridge University; and it is still in vogue among the Chinese fishermen, who find the skill of the cormorants more profitable than their own. Happily these birds are well cushioned with air spaces just under the skin to break the shock when they dive from a height and strike the water. The gluttony of a cormorant has passed into a proverb. It will continue to hunt every fish in sight, day after day, for its equally greedy masters, that only whet the bird's ravenous appetite from time to time, by removing its collar and allowing it to swallow an unenvied prize.

In some parts of the United States, but chiefly in the Bay of Fundy and beyond, the double-crested cormorants retire to nest in large companies on the ledges of cliffs along the sea, or in low bushes or bushy trees inland. The nest consists of a mass of sticks and sea-weed, and both it and its vicinity look as if they had been spattered over with whitewash, owing to the bird's unclean habits. When the four or six eggs are first laid, they are covered over with a rough, chalky deposit that is easily rubbed off, showing a bluish-green shell beneath. The young, that are hatched blind, have not even down to cover their inky-black skin. It takes fully two years to perfect the beautiful iridescent black plumage worn by adults. For a time the nestlings are fed with food brought up from their parents' stomachs; and so active is the cormorant's digestion that a fish caught by one is said to have reached a stage fit for baby food between the time the bird catches it in the water and transports it in its stomach to its adjacent nest. On shore these birds rest in an almost upright position, because their legs are set far back on their bodies, which also necessitates using the stiff tail as a prop. Doubtless this tail, that is used also as a rudder or paddle, adds to the cormorant's extraordinary facility in swimming under water.

LAMELLIROSTRAL, OR PLATE-BILLED SWIMMERS

Mergansers, or Fishing Ducks
River and Pond Ducks
Sea and Bay Ducks
Geese
Swans

LAMELLIROSTRAL, OR PLATE-BILLED SWIMMERS
(Order *Anseres*)

MERGANSERS; RIVER AND POND DUCKS; SEA AND BAY DUCKS; GEESE; SWANS
(Family *Anatidæ*)

Five subfamilies, numbering about two hundred species, constitute this large family of water fowl that in itself forms a well-defined order. They are the mergansers, river ducks, sea ducks, geese, and swans. All these birds have the margins of the beak *(rostrum)* furnished with lamels, or plates, tooth-like projections, fluted ridges or gutters along its sides; but the subfamilies are so well defined that their peculiarities would best be noted separately.

Mergansers, or Fishing Ducks
(Subfamily *Merginæ*)

Let the young housekeeper avoid any ducks with long, narrow, rounded, hooked, and saw-toothed bills; for the shelldrakes, or sawbills, as the mergansers are also called, have rank, unpalatable flesh, owing to their diet of fish, which are pursued and captured under water in the manner practiced by loons, cormorants, and other birds low in the evolutionary scale. Mergansers live in fresh as well as salt water.

American Merganser or Goosander.
Red-breasted Merganser.
Hooded Merganser.

Plate-billed Swimmers

River and Pond Ducks
(Subfamily Anatinæ)

The hind toe of these ducks is without a flap, or lobe, and the front of the foot is furnished with transverse scales, which are the two features of these birds which have led scientists to separate them into a distinct subfamily. But to even the untrained eye other peculiarities are also noticeable. The feet of these ducks are smaller than those of the sea-ducks, the toes and their webs naturally not being so highly developed, owing to the calmer waters on which they live; although some few species do associate with their sea-loving kin. They do not dive to pursue food like the mergansers and sea ducks, but nibble at the aquatic plants they live among, and dabble with their bills on the surface of the water for particles of animal matter; or, with head immersed and tail in air, probe the bottom of shallow waters for small mollusks, crustaceans, and roots of plants. Their bill acts as a sieve or strainer. From the more dainty character of their food, their flesh is superior. These drakes undergo a double moult; generally the sexes are distinct in color; the young resemble the female; but the wing-markings, in which a brilliant speculum is usually conspicuous, are the same in both sexes. When the males are not polygamous, they devote themselves to one mate, leaving the entire care of the young, however, to her. The speed of these ducks on the wing has been estimated anywhere from one hundred to a hundred and sixty miles an hour.

      Mallard Duck.
      Black or Dusky Duck.
      Gadwall, or Gray Duck.
      Baldpate, or Widgeon.
      Green-winged Teal.
      Blue-winged Teal.
      Shoveler.
      Pin-tail.
      Wood Duck.

Sea and Bay Ducks
(Subfamily Fuligulinæ)

The lobe, or web, hanging free on the hind toe is the characteristic looked for by scientists to separate these birds from the preceding group, the transverse scales on the front of the foot being common to both subfamilies. The toes and webs of these sea ducks are noticeably larger than those of the river ducks, owing to their greater exercise; and the feet are also placed a little farther back, which increases their facility in diving and swimming. Several of the species associate with the river ducks in still waters, the subfamily not being so exclusively maritime as its name would imply. Indeed, there seem to be notable exceptions to almost every general rule that might be applied to it except the one that relates to the formation of the toes. It is often said that the flesh of sea ducks, that feed more on mollusks, crustaceans, and other marine food, although not on fish, and less upon grain and other vegetable matter, is coarser, less palatable, and even sometimes inedible; but what of the canvasback duck, that peerless delicacy of the epicure?

> Red-headed Duck.
> Canvasback.
> Greater Scaup Duck, or Broadbill.
> Lesser Scaup, or Creek Broadbill.
> Ring-necked Duck.
> Golden-eye, or Whistler.
> Barrow's Golden-eye.
> Buffle-head, or Butter-ball.
> Old Squaw, or South Southerly.
> Harlequin Duck.
> American Eider Duck.
> King Eider.
> American Scoter, or Black Coot.
> White-winged Scoter.
> Surf Scoter.
> Ruddy Duck.

Geese

(Subfamily Anserinæ)

Cheeks, or lores, completely feathered where the swans are naked; tarsus, or lower part of leg, generally longer than the middle toe without the nail; scales on its front rounded: these are the purely scientific distinctions of the birds of this subfamily. Neck is midway in length between that of the ducks

Plate-billed Swimmers

and of the swans. Body is not so flat as the duck's and more elevated on the longer legs. Geese, that spend far more time on land, walk better than ducks, and depend altogether on a vegetable diet. When we see them tipping, with head immersed in the water and tail in air, they are probing the bottom for roots and seeds of plants, not for water insects or mollusks. In common with swans they resent intrusion by hissing with outstretched necks and by striking with the wings. When wounded on the water, a goose dives; then, with only its bill exposed above the surface, strikes out for land, where it evidently feels more at home. The sexes are generally alike in plumage, which undergoes only one moult a year; and both parents attend to the young as no self-respecting drake would do. A wedge-shaped flock of migrating geese, with an old gander in the lead at the point of the V, old sportsmen say, is a familiar sight in the spring and autumn skies, that echo with the *honk, honk*, or noisy cacklings, coming from the distended necks of the travellers.

White-fronted Goose.
Snow Goose.
Lesser Snow Goose.
Canada, or Wild Goose.
Brant.
Black Brant.

Swans

(Subfamily Cygninæ)

Bare skin between the eye and bill is the scientific mark of distinction between swans and geese; many other points of difference are too well known to mention. Swans feed on small mollusks in addition to vegetable matter which they secure by "tipping" or by simply immersing their long, graceful necks. They migrate in V-shaped flocks like the geese, and often utter loud, trumpeting notes unlike the noisy gabble of both geese and ducks. Plumage of sexes alike.

Whistling Swan.
Trumpeter Swan.

MERGANSERS, OR FISHING DUCKS

(Subfamily Merginæ)

American Merganser
(Merganser americanus)

Called also: GOOSANDER; SHELLDRAKE; SAW-BILL; FISHING DUCK; DIVING GOOSE; BUFF-BREASTED SHELLDRAKE; WEASER; DUN DIVER.

Length—23 to 27 inches.

Male—Head, which is slightly crested, and upper neck, glossy greenish black; hind neck, breast, and markings on wings, white; underneath delicately tinted with salmon buff. Back black, fading to ashy gray on the lower part and tail. Wings largely white; tips of the coverts white, forming a mirror, and banded with black. Bill toothed and red, or nearly so, and with black hook, and nostrils near the middle.

Female and Young—Smaller than male; head and upper neck reddish brown; rest of upper parts and tail ashy gray; breast and underneath white.

Range—North America generally, nesting from Minnesota northward, and wintering from New England, Illinois, and Kansas southward to southern States.

Season—Winter resident from November to April.

A surprising number of popular names have attached themselves to this large, handsome swimmer that studiously avoids populated regions and the sight of man; that no sportsman would, or, indeed could, eat; that eludes pursuit by some very remarkable diving and swimming feats, and therefore enjoys popularity in names alone. Its preferences are for remote waterways at the north, where its family life is spent, only a few nests being reported this side of the Canadian border; but when a hard crust of ice locks up their fish, frogs, mollusks, and other aquatic animal food, small companies of six or eight mergansers migrate

Mergansers

to our lakes, rivers, and the ocean shore to hunt there until spring. Salt and fresh water are equally enjoyed.

Feeding appears to be the chief object in life of this gluttonous bird that often swallows a fish too large to descend entire into the stomach, and must remain in the distended throat until digested piecemeal. Its saw-like bill for holding slippery prey, and rough tongue covered with incurved projections like a cat's, doubtless help speed the process of digestion, which is so rapid as to keep the bird in a constant state of hunger, and drive it to desperate rashness to secure its dinner. It will plunge beneath a rushing torrent after a fish, or dive to great depths to secure it, swimming under water with long and splendidly powerful, dexterous strokes that soon overtake the fish in its own element. These feats, with the sudden dropping out of sight practiced so artfully by the loons, make a merganser an exceedingly difficult mark for the sportsman to hit; and its muscular, tough, rank flesh offers no reward for his efforts. Usually these birds depend upon the water to escape danger; but when disturbed in a shallow fishing ground, a flock seems to run along the water for a few yards, patting it with their strongly webbed feet, then rising to windward, they head off in straight, strong, and rapid flight, toward distant shelter.

The adult male in his nuptial dress is a conspicuously beautiful fellow, with his dark, glossy green head, rich salmon-colored breast, and black and white wings, set off by a black back. But this attire is not worn until maturity, in the second year; and in the intervening time, as well as after the nesting season is over, he looks much like his mate and their young. Birds whose upper parts show the grayish brown that predominates when we see them in winter are called "dun divers" in many sections. It is the male bird in spring plumage that the taxidermist mounts to decorate the walls of dining-rooms and shooting lodges.

Mergansers build a nest of leaves, grasses, and moss, lined with down from their breasts, in a hole of a tree or cliff, where from six to ten creamy-buff eggs are laid in June, and tended exclusively by the mother, even after they have evolved into fluffy ducklings. At this time the drake is undergoing a thorough moult.

RED-BREASTED MERGANSER
½ Life size.

Red-breasted Merganser
(Merganser Serrator)

Called also:—SHELLDRAKE; SAWBILL; WHISTLER; PIED SHELLDRAKE; GARBILL

Length—22 to 24 inches.

Male—Head and throat greenish black; more greenish above, and with long, pointed crest over top of head and nape; white collar around neck; sides of lower neck and the upper breast cinnamon red, with black streaks; lower breast, underneath, and the greater part of wings white; other feathers black. Back black; lower back and sides finely barred with black and white; a white patch of feathers, with black border, in front of wings, and two black bars across them. Bill long, saw-toothed, red, curved at end, and with nostrils near the base; eyes red; legs and toes reddish orange.

Female and Young—Similar to the American merganser. Head, neck, and crest dull, rusty brown; dark ashy on back and tail; throat and under parts white, shaded with gray along sides; white of wing restricted to a patch (mirror or speculum); no peculiar feathers in front of wing.

Range—United States generally; nests from Illinois and Maine northward to Arctic regions; winters south of its nesting limits to Cuba.

Season—Winter resident and visitor; October to April.

Swift currents of water, deep pools where the fish hide, and foaming cataracts where they leap, invite the red-breasted merganser, as they do its larger American relative; for both birds have insatiable appetites, happily united with marvelous swimming and diving powers that must be constantly exercised in pursuit of their finny prey. Fish they must and will have, in addition to frogs, little lizards, mollusks, and small shell fish; and for such a diet this fishing duck forsakes its northern nesting grounds in winter, when ice locks its larder, to hunt in the open waters, salt or fresh, of the United States. Cold has no terror for these hardy creatures; they swim as nimbly in the icy water of the St. Lawrence as in the rivers of Cuba, and disappear under an ice cake with no less readiness than they do under lily-pads. Food is their chief desire; and rather than let a six-inch fish go, any merganser would choke in its efforts to bolt it.

Mergansers

Their appetite is so voracious that often some of their food must be disgorged from their distended crops before the birds are able to rise from the water. An almost exclusive fish diet, with the constant exercise they must keep up to secure it, makes their flesh so rank and tough that no sportsman thinks of shooting the mergansers for food; and by sudden, skilful dives the birds are as difficult to kill as the true "water witches." Only the youngest, most inexperienced housekeeper thinks of buying any saw-billed duck in market; the serrated edges indicating that the bill is used as a fish chopper, and fish food never makes flesh that is acceptable to a fastidious palate.

In the United States, at least, the red-breasted mergansers are far more abundant than the preceding species, which they very closely resemble after the nuptial dress has been laid aside for the brown and gray winter plumage. Males may be distinguished by the color of their breasts at any time; but the females and young of both species are most bewilderingly similar at a little distance. The position of the nostril, near the centre of the American merganser's bill, and near the base of the red-breasted species, is the positive clew to identity. The latter bird's croak is another aid. All mergansers look as if they needed to have their hair brushed.

While the construction of the nest of these sometimes confused relatives is the same, the red-breasted merganser makes its cradle directly on the ground, among rocks or bushes, but never far from water. It is the female that bears all the burden of hatching the creamy buff eggs—six to twelve—and of feeding and training the young brood; her gorgeous, selfish mate discreetly withdrawing from her neighborhood when nursery duties commence. But the long-suffering mother bird is a perfect pattern of all the domestic virtues. "I paddled after a brood one hot summer's day," says Chamberlain, "and though several times they were almost within reach of my landing net, they eluded every effort to capture them. Throughout the chase the mother kept close to the young birds, and several times swam across the bow of the canoe in her efforts to draw my attention from the brood and to offer herself as a sacrifice for their escape."

Hooded Merganser

(Lophodytes cucullatus)

Called also: HAIRY HEAD; WATER PHEASANT; HOODED SHELLDRAKE

Length—17 to 19 inches.

Male—Handsome semicircular black crest with fan shaped patch of white on each side of greenish black head; upper parts black, changing to brown on lower back; lower fore neck, wing linings, and underneath white, finely waved with brownish red, and dusky on sides. Two crescent shaped bands of black on sides of breast. A white speculum or mirror on wing, crossed by two black bars. Bill bluish black, with nostrils in basal half; eyes yellow.

Female—Smaller; dark ashy brown above, minutely barred with black; more restricted and reddish brown crest, lacking the white fan; under parts white; sides grayish brown.

Young—Similar to female, but without crest; no black and white bars before wing; wings scarcely showing the white mirror.

Range—North America; nests throughout its range; winters in southern United States, also in Cuba and Mexico.

Season—Chiefly a winter resident and visitor south of the Great Lakes and New England.

Unlike the two larger mergansers that delight in rushing torrents and in making daring plunges beneath them, this strikingly beautiful "water pheasant," as it is sometimes called, chooses still waters, quiet lakes and mill-ponds for a more leisurely hunt after small fish, mollusks, and water insects, adding to this menu roots of aquatic plants, seeds, and grain. It is claimed that this variation in the fish diet, and the consequent lack of hardening of the muscles, make the merganser's flesh edible; and in spite of its saw-toothed bill, the certain index of rank, fishy flesh, epicures insist that this is an excellent table duck; but in just what state of rawness it is most delicious, who but an epicure may say?

"It seems an undue strain on the imagination, not to say palate, to claim that any of the fish-eating ducks are edible," says Mr. Shields. "Men who kill everything they can find in the woods, in the fields, or on the water, say all mergansers, coots and grebes are good if properly cooked. When asked what this proper method of cooking is, they say the birds should first be parboiled through two or three waters; that they should then be well baked, stewed, fricassed, or broiled, and flavored with rashers of bacon and onions, potatoes, etc. This means, then, that the bird should be so treated as to rob it of all its original quality, and to reduce it to a condition simply of meat. A hawk, an owl, a cayote, a catfish, a German carp, or even a dogfish may be made edible by such treatment. If a bird or a fish is not fit to eat without all this manipulation and seasoning, it is not an edible animal in the first place. Then why kill it?"

Like the wood duck, golden-eye, bufflehead, and its immediate kin, the hooded merganser goes into a hollow tree or stump to build a nest of grasses, leaves, and moss, lined with down from the mother's breast, and lays from eight to ten buffy white eggs. Now is the time that the handsome male disports himself at leisure, and at a distance, while the patient little mother keeps the eggs warm, feeds the yellowish nestlings, carries them to the lake one by one in her bill, as a cat carries its kittens; teaches them to swim, dive, and gather their own food, and to fly by midsummer; defends them with her life, if need be; and welcomes home the lazy, cavalier father when the drudgeries are ended and the young are fully able to join the migrating flocks that begin to gather in the Hudson Bay region in September. It is she who ought to wear the white halo around her head instead of the drake.

Sportsmen often find small companies of hooded mergansers in the same lake with mallard, black, wood, and other ducks that, like them, delight in woody, well-watered interior districts. Mr. Frank Chapman found them in small ponds in the hummocks of Florida; and the author first made their acquaintance on a poultry stand in the French market in New Orleans.

RIVER AND POND DUCKS
(*Subfamily Anatinæ*)

Mallard Duck
(*Anas boschas*)

Called also: WILD OR DOMESTIC DUCK; GREEN HEAD

Length—23 inches.

Male—Head and neck glossy green with white ring like a collar defining the dividing line from the rich chestnut breast; underneath grayish white, finely marked with waving black lines; back dark grayish brown, shading to black on lower back and tail. Four black upper feathers of tail curve backward; rest of tail white, black below. Speculum or wing-bar rich purple with green reflections and bordered by black and white. Bill greenish yellow with gutters on the side.

Female—Plumage generally dark brown varied with buff; breast and underneath buff, mottled with grayish brown; wings marked like male's.

Range—Nests rarely from Indiana and Iowa and chiefly from Labrador northward; winters from Chesapeake Bay and Kansas southward to Central America. Rare in New England.

Season—Winter resident in southern states; a transient visitor or migrant, during the winter months, at the north.

Small, grassy ponds, slow-moving streams, sloughs, and the labyrinths of lakes and rivers that are thickly grown with wild rice and rushes, such as abound in the interior of the United States and Canada, make the ideal resort of the mallards, or, indeed, of most ducks dear to the sportsman's heart. Here large companies gather in August and September when the ripened grain invites them to the feast they most enjoy, flying at dusk or by night in wedge-shaped battalions from their resting-grounds at the far north, to remain until the ice locks up their food and they must shift their home farther south. In Illinois,

River and Pond Ducks

Minnesota, Iowa, and Indiana, they are among the first ducks to arrive and the last to leave with the hardy scaups or bluebills. And in sheltered localities a few sometimes winter, just as a few break through traditions and nest in secluded spots in the same states; but from Kansas and the Chesapeake country southward, they may be positively relied upon until the time arrives for the spring migration, however more abundant they may be in the interior than along our coast. Let no one imagine that because some ducks are classified in the books as "river and pond," and others as "sea and bay ducks," they are not often found in the same places. It is the lobed hind toe of the latter group that really differentiates them, and not always their habitats.

Well concealed in the tall sedges that literally drop food into their gaping mouths, the mallards feed silently upon the ripe grain and seeds, dabbling on the surface of the water or, suddenly tipping tail upward and stretching head downward in the shallow waters, probe the muddy bottom for the small mollusks, fish, worms, rootlets, and vegetable matter they delight in. When a good mouthful has been taken in the bill is closed tight, thus forcing out through the gutters along the sides, that act as strainers, the mud and water that were taken in with the food. Ripe corn that has dropped in the fields is a favorite cereal. Fish and animal substances form a small fraction of the mallards' diet; they are very near to being vegetarians, the fact that makes their flesh so delicious.

"In the spring and fall the Kankakee region of Illinois and Indiana is one of the finest grounds for mallards, teal, wood-duck and geese, to be found in the United States," says Maurice Thompson. "I need not say to the sportsman that the mallard is the king's own duck for the table. The canvasback does not surpass it. I have shot corn-fed mallards whose flesh was as sweet as that of a young quail, and at the same time as choice as that of the woodcock."

Instead of becoming indolent and moody after a plentiful dinner, these ducks are uncommonly lively. They jabber among themselves, spatter the water freely, half fly, half run along the surface of the lake, and are positively playful so long as the leader of the sport, that is on the constant lookout, gives no sign of warning. One might think they were mad, but often their frantic antics indicate that insects are troubling them, and all their splut-

tering and diving is done to get rid of the pests. Mallards dive and swim under water also to escape danger, but rarely to collect food. During the day they make many bold excursions to the centre of the lake and explore the inlets and indentations of the shore. On the first *quack* of alarm, however, up bounds the entire flock and, rising obliquely to a good height, their stiffened wings whistling through the wind, off they fly at a speed no locomotive can match. Perhaps the reason for most misses of the amateur hunter is his inability to conceive the rate at which ducks move, and so to hold far enough ahead of the bird he has selected. Mallards waste no time sailing, but after climbing the sky on throbbing wings they continue to flap them constantly. Before alighting it is their habit to wheel round and round a feeding-ground to assure themselves no danger lurks in ambush. They are conspicuous sufferers from the duckhawk, whose marvelous flight so far excels even theirs that escape is hopeless in a long race unless the duck should be flying over water, into which a sudden plunge and a long swim under the surface to a sheltered corner in the sedges, frees it from the persecutor that lives by tearing the flesh from the breasts of hundreds of such victims every year.

Wary as these ducks are, they are also eminently inquisitive, or the painted, wooden decoys of dingy little females, gay bandana handkerchiefs fluttering from poles, that are used in the south to excite their curiosity, and other time-honored tricks of sportsmen would never have been crowned with success. The mallards are also exceedingly shy, and feel at greatest ease and liberty when the dusk of evening and dawn covers their feeding-grounds and conceals their flight that is often suspected solely by the whistling of their wings through the darkness overhead. Their loud *quack, quack*, exactly like that of the domestic duck, resounds cheerfully in the spring and autumn migrations.

To see the endearments and little gallantries the handsome drake bestows on his mate in spring, no one would suspect him of total indifference to her later. Waterton and other writers claim that the wild mallard is not only strictly monogamous, but remains paired for life. Perhaps polygamy cannot be fairly charged against him, however suspicious his indifference to his mate and ducklings appears. Many ornithologists claim that he is positively unable to help his mate and young, owing to the extra molt his plumage undergoes at the end of June, when he actually loses the power of flight for a time and does not regain his beautiful full plumage until the autumn. But certainly the character of the domesticated mallard must have sadly deteriorated, if this is so, for in the barn-yard, at least, he is a veritable Mormon.

In a nest lined with down from her breast, and made of hay, leaves, or any material that can be scraped together on the ground, near the water or in a bushy field back from it, the mother confines herself for twenty-eight days. It is then her gay cavalier goes off to his club, or its equivalent, with other like-minded pleasure-seekers, while she bears the full burden of the household. Very seldom does she leave the pale bluish or greenish gray eggs—six to a dozen—to get food and a brief swim in the lake; and she is careful to pull the down coverlet well over the eggs to retain their heat during her outings. As her incubating duties near their end, she usually does not stir from the nest at all. There are some few records of nests made in trees. If the nest is near the water, on the ground, the young ones instantly make for it when they leave the shell; but being unable to walk well at first, the overworked mother must carry them to it in her bill, it is said, if the nest is far back on a bank. Many pathetic stories are in circulation, showing the mother's total self-forgetfulness and voluntary offering of her own life to protect the downy brood. Water-rats and large pike, that eat her babies when they make their earliest dives, are the worst enemies she has to fear until they are able to fly, some six weeks or more after hatching, and the duck-hawk finds them easy prey.

The mallard is by far the most important species we have, as it is the most plentiful, the most widely distributed, and the best known, being the ancestor of the common domestic duck; and although many of its habits have undergone a change in the poultry-yards, others may still be profitably studied there by those unable to reach the inaccessible sloughs, bayous, and lagoons where the wild ducks hide.

Black Duck

(Anas obscura)

Called also: DUSKY DUCK ; DUSKY MALLARD

Length—22 to 23 inches; same size as the mallard.

Male and Female—Resembling the female mallard, but darker and without white anywhere except on the wing linings; violet blue patch or speculum on wings bordered by black—a fine white line on that of male only. General plumage dusky brown, not black, lighter underneath than on upper parts, the feathers edged with rusty brown. Top of head rich, dark ashy brown, slightly streaked with buff; sides of head and throat pale buff, thickly streaked with black. Female paler yellow. Bill greenish. Feet red.

Range—"Eastern North America, west to the Mississippi Valley, north to Labrador, breeding southward to the northern parts of the United States."—A. O. U.

Season—Resident in the United States, where it nests; also winter resident, from September to May; most abundant in spring and autumn migrations.

In New England and along the Atlantic States, where the mallard is scarce, the black duck (which is not black but a dusky brown), replaces it in the salt-creeks and marshes as well as on the inland rivers, lakes, and ponds; and even the sea itself is sometimes sought as an asylum from the gunners. Not all river and pond ducks confine themselves to the habitats laid down for them in the books. Black ducks, when persistently hunted, frequently spend their days on the ocean, returning to their favorite lakes and marshes under cover of darkness—for they are exceedingly shy and wary—to feed upon the seeds of sedges, corn in the farmer's fields, the roots and foliage of aquatic plants, and other vegetable diet, which is responsible for the delicious quality of their flesh, so eagerly sought after.

Brush-houses thatched with sedges, that are set up in the duck's feeding-grounds by hunters, may not be distinguished from the growing plants in the twilight or early dawn ; wooden decoys easily deceive the inquisitive birds; live domestic ducks tied by the leg to the shore, though apparently free to swim at large, lure the wild ones near the gunners in ambush, and numerous other devices, long in vogue among men who spare themselves the fatigue of walking through the sedges to flush their victims, help pile the poultry stalls of our city markets just as soon as the law allows in autumn. In the early spring, when the law is still "open" and should be closed, housekeepers find eggs already well formed in this and other game birds brought to their kitchens. Of all the wild fowl that enter the United States, this duck, it is said, possesses the greatest economic value, which should be a sufficient reason, if no higher motive prompted, to give it the fullest protection. While the nesting season is from the last of April to the early part of June, the birds have mated many weeks before. They are the spring laws that need serious going over by our legislators.

So closely resembling the mallard in habits that an account of them need not be repeated here, the black duck is not so common in the interior nor in the south, for it was the Florida duck that early ornithologists confounded with this species, which, they claimed, had the phenomenal nesting range extending from Labrador to the Gulf. Illinois and New Jersey are as far south as its nests have been found. The black duck, that seems to have a more hardy constitution than many of its kin, stays around our larger ponds long after the ice has formed, and where springs keep open pools, it is not infrequently met with all through a mild winter.

Gadwall

(*Anas strepera*)

Called also: GRAY DUCK

Length—20 to 22 inches.

Male—Upper parts have general appearance of brownish gray, waved and marked with crescent-shaped white and blackish bars. Top of head streaked with black or reddish brown; sides of head and neck pale buff brown, mottled with darker; lower neck and breast black or very dark gray, each feather marked with white and resembling scales; grayish and white underneath, minutely lined with gray waves; lower back dusky, changing to black on tail coverts; space under tail black. Wings chestnut brown, gray, and black, with white patch framed in velvety black and chestnut. Wing-linings white. Bill lead color. Feet orange.

Female—Smaller than male and darker. Head and throat like male's; back dark grayish brown, the feathers edged with

buff; breast and sides buff, thickly spotted with black, but the female throughout lacks the beautiful waves, scales, and crescent-shaped marks that adorn her mate. Underneath, including under tail-coverts and wing-linings, white. Little or no chestnut on wings; speculum or wing-patch white and gray. Bill dusky, blotched with orange. Immature birds resemble the mother.

Range—Cosmopolitan; nests in North America, from the middle states northward to the fur countries, but chiefly within United States limits. Most abundant in Mississippi Valley region and west; also northward to the Saskatchewan.

Season—Winter resident south of Virginia and southern Illinois; winter visitor, most abundant in spring and autumn migrations, north of Washington.

This beautiful species, first discovered by Wilson, on the shores of Seneca Lake, New York, keeps close by fresh water, showing no liking whatever for the sea as the black duck does. In the Atlantic states the gadwall is rare, except as a migratory visitor inland, while in the sloughs of the Mississippi Valley, Florida, and the Gulf states, it is abundant in favored spots that other ducks frequent when the wild rice and field-corn ripen, and that local sportsmen also revel in. The gadwall's flesh is particularly fine; its mixed diet of grain and small aquatic animal food imparting a gamy flavor to it that epicures appreciate.

As this duck is very shy and full of fear, it dozes most of its time away when the sun is high, securely hidden in the tall sedges that line the marshy lake or quiet stream; and emerging at twilight to feed, to disport itself with its companions, to lift up its voice in happy bubblings and quacks, to fly from lake to lake in wedge-shaped companies, it pursues, under cover of partial or even total darkness, the round of pleasures and duties customary among all the duck tribe. In nesting and other habits as well, the gadwall so closely resembles the mallard that a description of them would be merely a repetition. Even its voice is very like the mallard's, although the *quack* is more frequently repeated; but Gesner must have discovered some unusually shrill, high-pitched notes in it when he added *strepera* to the bird's name.

Baldpate

(Anas americana)

Called also: AMERICAN WIDGEON

Length—18 to 20 inches.

Male—Crown of head white or buff, sides of head, from the eye to the nape, have broad band of glossy green, more or less sprinkled with black; cheeks and throat buff, marked with fine lines and bars of black; **upper breast** and sides light reddish, violet brown (vinaceous), **each feather** with grayish edge forming bars across breast. More **grayish** sides **are** finely waved with black; lower parts **and** wing-linings white; black under tail. Back grayish brown, more or less tinged with the same color as breast, and finely marked with black. Wings have glossy green patch bordered by velvety black. Bill grayish blue with black tip. Feet and legs dusky.

Female—Smaller. Head and throat white or cream, finely barred with black and without green bands; darker above; upper breast and sides pale violet, reddish brown washed with grayish, interrupted with whitish or gray bars. Wings like male's, though the speculum may be indistinct and gray replace the white; back grayish brown, **the feathers** barred with buff.

Range—North America; **nests regularly from** Minnesota northward, and casually as far as **Texas**, but not on the Atlantic coast. Winters in the United **States,** from southern states to the Gulf; also in Guatemala, **Cuba**, and northern South America.

Season—Spring and **autumn visitor,** and winter resident, October to April.

The baldpates, keeping just in advance of the teeth of **winter** with the large army of other ducks that come flying out of the north in wedge-shaped battalions when the first ice **begins** to form, break their long journey to the Gulf states and the tropics by a prolonged feast in the wild rice, sedges, and celery in northern waters, both inland and along the coast. A warm reception of hot shot usually awaits them all along the line, for when celery-fed or fattened on rice their **flesh can** scarcely be distinguished from that of the canvasback duck, and sportsmen and pot-hunters exhaust all known devices to lure them within gun-range. The gentleman hidden behind "blinds" on the "duck-shores" of

BALD-PATE DUCK.
½ Life-size.

Maryland and the sloughs of the interior, and with a flock of wooden decoys floating near by; or the nefarious market-gunner in his "sink boat," and with a dazzling reflector behind the naphtha lamp on the front of his scow, bag by fair means and foul immense numbers of baldpates every season; yet so prolific is the bird, and so widely distributed over this continent, that there still remain widgeons to shoot. That is the fact one must marvel at when one gazes on the results of a single night's slaughtering in the Chesapeake country. The pot hunter who uses a reflector to fascinate the flocks of ducks that, bedded for the night, swim blindly up to the sides of the boat, moving silently among them, often kills from twenty to thirty at a shot. True sportsmen must soon awaken to the necessity for stopping this wholesale murdering of our finest game birds.

Whew, whew, whew—" a shrilly feeble whistle, precisely such as the young puddle duck of the barnyard makes in his earliest vocal efforts"—announces the coming of a flock of baldpates high overhead. Audubon heard them say "*sweet, sweet,*" as if piped by a flute or hautboy. In spite of their marvelously swift flight, estimated from one hundred to one hundred and twenty miles an hour, their stiffened wings constantly beating the air that whistles by them, they are, nevertheless, often overtaken on the wing by the duck hawk, their worst enemy next to man. Diving and swimming under water are their only resorts when this villain attacks them.

But when living an undisturbed life, the widgeons greatly prefer that other ducks, notably the canvasbacks, should do their diving for them. Around the Chesapeake, where great flocks of wild ducks congregate to feed on the wild celery, the widgeons show a not disinterested sociability, for they kindly permit their friends to make the plunges down into the celery beds, loosen the tender roots, and bring a succulent dinner to the surface; then rob them immediately on their reappearance. Such piracy keeps the ducks in a state of restless excitement, which is further induced by the whistling of the widgeons' wings in their confused manner of flight in and around the feeding-grounds. Here they wheel about in the air; splash and splutter the water; stand up in it and work their wings; half run, half fly along the surface, and in many disturbing ways make themselves a nuisance to the hunter in ambush. They seem especially alert and lively. Neither are they so shy as many of their companions; for when come upon suddenly in the coves of the lake, they usually row boldly out toward the centre, out of gun range, and take to wing, if need be, rather than spend their whole day dozing in the tall grasses on the shores as many others do. Not that they may never be caught napping on the sand flats or in the sedges when the sun is high, for all ducks show decided nocturnal preferences; only widgeons are perhaps the boldest of their associates. Open rivers, lakes, estuaries of large streams, and bays of the smaller bodies of salt water attract them rather than the sluggish, choked-up sloughs that shyer birds delight to hide in.

Instead of nesting close beside the water in the sedges, after the approved duck method, the widgeons commonly go to high, dry ground to lay from seven to twelve buff-white eggs in a mere depression among the leaves that the mother lines with down from her breast. Nests are frequently found half a mile or more from water. It is supposed, but not as yet proved, that the mother carries in her bill each tiny duckling to the water, where it is at home long before it feels so on land or in the air. At various stages of the bird's development the plumage undergoes many changes; but aside from those of age and sex, the baldpates show unusual variability. However, Dr. Coues consoles the novice with the assurance that "the bird cannot be mistaken under any conditions; the extensive white of the under parts and wings is recognizable at gunshot range."

. . . .

The European Widgeon (*Anas penelope*) has found its way across the Atlantic and our continent, for it nests in the Aleutian Islands as well as in the northern parts of our eastern coast. It is occasionally met with in the eastern United States; and, although it has a bald pate also, its blackish throat and the reddish brown on the rest of the head and neck easily distinguish it from its American prototype.

Green-winged Teal
(*Anas carolinensis*)

Length—14 inches. One of the smallest ducks.

Male—Head and neck rich chestnut, with a broad band of glossy green running from eyes to nape of neck; chin black; breast light pinkish-brown, spotted with black; upper back and sides finely marked with waving black and white lines; lower back dark grayish brown, underneath white. A white crescent in front of the bend of the wing; wings dull gray, tipped with buff and with patch or speculum half purplish black and half rich green. Head slightly subcrested. Bill black. Feet bluish gray.

Female—Less green on wings; no crest; throat white; head and neck streaked with light reddish brown on dark-brown ground; mottled brownish and buff above; lower parts whitish changing to buff on breast and lower neck, which are clouded with dusky spots.

Range—North America at large; nests in Montana, Minnesota, and other northern states, but chiefly north of the United States; winters from Virginia and Kansas, south to Cuba, Honduras, and Mexico.

Season—Spring and autumn migratory visitor north of Washington and Kansas; more abundant in the interior than on the coasts.

Next to the wood duck, this diminutive, exquisitely marked and colored kinsman is perhaps the handsomest member of its tribe; and, next to the merganser, it is said to be the most fleet of wing as it is of foot, unlike many of its waddling relations; but epicures declare its delicious flesh is the one characteristic worth expending superlatives upon. When the teal has fed on wild oats in the west, or on soaked rice in the fields of Georgia and Carolina, Audubon declared it is much superior to the glorified canvasback. Nothing about its rankness of flavor when it has gorged on putrid salmon lying in the creeks in the northwest, or the maggots they contain, ever creeps into the books; and yet this dainty little exquisite of the southern rice fields has a voracious appetite worthy of the mallard, around the salmon canneries of British Columbia, where the stench from a flock of teals passing overhead betrays a taste for high living, no other gourmand can approve. When clean fed, however, there is no better table-duck than a teal.

River and Pond Ducks

Among the earliest arrivals from the horde of water-fowl that follow the food supply from the far north into the United States every autumn, the green-wings are exceedingly abundant in the fresh water lakes and ponds of the interior, and less so on the salt water lagoons and creeks of the coast until frost locks up the celery, sedges, wild rice, berries, seeds of grasses, tadpoles, and the various kinds of insects on which they commonly feed. Then the teals go into winter quarters, and as they pass in small, densely packed companies overhead, the peculiar reed-like whistling of their swift wings may be plainly heard. Old sportsmen tell of clouds of ducks, numbering countless thousands, but they best know why such flights are gone forever from the United States.

The selfish, dandified drakes, that have spent their summer putting on an extra suit of handsome feathers and living an idle life of pleasure while their mates attended to all the nursery duties, leave them to find their way south as best they may, while they pursue a separate course. In the spring the teals are, perhaps, the easiest ducks to decoy. To watch the gallantries and antics of the drake in the spring, when he proudly swims round and round his coy little sweetheart, uttering his soft whistle of endearment, no one would accuse him of total indifference to her later. Happily, she is self reliant, dutiful to her young, courageous, resourceful. As a brood may consist of from six to sixteen ducklings, the mother does not lack company during the autumn migration, though she must often pay heavy toll to the gunners in every state she passes through. Were she not among the most prolific of birds, doubtless the species would be extinct to-day. Happily this duck is a mark for experts only; for, with a spring from the water, it is at once launched in the air on a flight so rapid that few sportsmen reckon it correctly in taking aim. When wounded, the teal plunges below the water, or when pursued by a hawk; but it rarely, if ever, dives for food, the "tipping-up" process of securing roots of water plants in shallow waters answering the purpose. Occasionally one sees a flock of teals sunning themselves on sandy flats and bogs, preening their feathers, or dozing in the heat of noon; then the hunter picks them off by the dozen at a time; but ordinarily these birds keep well screened in the grasses at the edges of the waters until twilight. While, like most other ducks, they are

GREEN-WINGED TEAL.

particularly active toward night and at dawn, they are not so shy as many. Farmers often see them picking up corn thrown about the barnyard; and Mr. Arnold tells in the "Nidologist" of finding nests of the green-winged teals built in tufts of grass on the sun baked banks along the railroad tracks in Manitoba, where the workmen constantly passed the brooding females intent only on keeping warm their large nestful of cream-white eggs. Nests have been found elsewhere, quite a distance from water, which would seem scarcely intelligent were not the teals very good walkers from the first, and less dependent than others on the food water supplies. In the west one sometimes surprises a brood and its devoted little mother poking about in the undergrowth for acorns, or for grapes, corn, wheat, and oats that lie about the cultivated lands at harvest time. Green-wings are early nesters, and have full fledged young in July, when the blue-wings and cinnamon teal are still sitting.

Blue-winged Teal

(*Anas discors*)

Called also: WHITE-FACED TEAL; SUMMER TEAL

Length—15 to 16 inches.

Male—Head and neck deep gray or lead color with purplish reflections; black on top; a broad white crescent bordered by black in front of head; breast and underneath pale reddish buff, spotted with dusky gray on the former and barred on the flanks. Back reddish brown, marked with black and buff crescents, more greenish near the tail. Shoulders dull sky blue; wing patch green bordered with white. Bill grayish black. Feet yellowish with dusky webs.

Female—Dusky brown marked with buff, with an indistinct white patch on chin; sides of the head and neck whitish, finely marked with black spots except on throat; breast and underneath paler than male in winter; wings similar but with less white. In summer plumage males and females closely resemble each other.

Range—North America from Alaska and the British fur countries to Lower California, the West Indies, and South America; nests from Kansas northward; winters from Virginia and the lower Mississippi Valley southward. Most abundant east of the Rocky Mountains.

River and Pond Ducks

Season—More common in the autumn migrations, August, September, and October, along the Atlantic coast states than in the spring, and always more plentiful in the Mississippi region than near salt water.

Similar in most of its habits to the green-winged teal, the blue-winged species appears a trifle less hardy, and is therefore, perhaps, the very first duck to come into the United States in the early autumn and to hurry southward when the first frost pinches. Tropical winters suit it perfectly, but many birds remain in our southern states until spring. Here they forget family traditions of shyness, when the sun shines brightly, and sit crowded together basking in its rays on the mud flats and shallow lagoons, delighting in the tropical warmth. It is when they are enjoying such a sun bath that the pot hunter, who has stolen silently upon them, discharges an ounce of shot in their midst, and bags more ducks at a time than one who knows how scarce this fine game bird is, where once it was exceedingly abundant, cares to contemplate. The old "figure four" traps, to which ducks are decoyed with rice, still find favor with the market hunter, who is looking for large returns for his efforts, rather than for sport. Decoys are all but useless in autumn when the drakes show no attention to even their mates.

Formerly these teals were very common indeed in New England, the middle Atlantic and the middle states, whereas for many seasons past the same old story is heard there from the sportsmen: "There is a very poor flight this year." It is likely to grow poorer and poorer in future unless the ducks are given better protection. We must now go to the inaccessible sloughs, grown with wild rice, in Minnesota, Illinois, Indiana, and westward, or to the lagoons of the lower Mississippi Valley to find the two commoner species of teals in abundance. In such luxuriant feeding-grounds, where they associate closely, long, wedge-shaped strings of ducks rise from the sedges at any slight alarm, and shoot through the air overhead on whistling wings. We are accustomed to seeing small, densely massed flocks in the east when the birds are migrating southward. The blue-winged teals, after their small size is noted, can always be distinguished by the white crescent between the bill and eyes, conspicuous at a good distance. "When they alight, they drop down suddenly among the reeds in the manner of the snipe or woodcock," says

Nuttall, instead of hovering suspiciously over the spot for awhile, like the mallards. They are silent birds, and, though not always actually so, their low, feeble *quack*, rapidly repeated, is so diminutive that they get little credit for a vocal performance.

Shoveler

(*Spatula clypeata*)

Called also: SPOONBILL; BROADBILL

Length—18 to 20 inches.

Male—Head and neck dusky, glossy bluish green; back brown, paler on the edges of the feathers, and black on lower back and tail; patches on sides of base of tail, lower neck, upper breast, and some wing feathers white; lower breast and underneath reddish chestnut; shoulders grayish blue; wing patch green. Bill longer than head, twice as wide at end as at base, and rounded over like a spoon; teeth at the sides in long, slender plates. Tail short, consisting of fourteen sharply pointed feathers. Feet small and red.

Female—Smaller, darker, and duller than male. Head and neck streaked with buff, brown, and black; throat yellowish white; back dark olive brown, the feathers lighter on the edges; underparts yellowish brown indistinctly barred with dusky; wings much like male's, only less vivid. Immature birds have plumage intermediate between their parents'; their shoulders are slaty gray and the wing patch shows little or no green.

Range—"Northern hemisphere; in America more common in the interior; breeds regularly from Minnesota northward and locally as far south as Texas; not known to breed in the Atlantic States; winters from southern Illinois and Virginia southward to northern South America." (Chapman.)

Season—Winter visitor in the south; spring and autumn migrant north of Washington; more abundant in autumn migrations in the east.

However variable the plumage of this duck may be in the sexes and at different seasons, its strangely shaped bill at once identifies it, no other representatives of the spoonbill genus of ducks having found their way to North American waters. Apparently the shoveler is guided by touch rather than sight, as it pokes about on the muddy shores of ponds or tips up to probe in the shallow waters for the small shellfish, insects, roots of aquatic plants, and small fish it feeds on. It is not a strict vegetarian, however delicate and delicious its flesh may be at the proper season. There are many sportsmen who would not pass a shoveler to shoot a canvasback.

North of the United States, where these ducks chiefly have their summer home, we hear of the jaunty, parti-colored drake, gayly decked out for the nesting season, when he is truly beautiful to behold, and charmingly attentive to his more sombre mate. By the time the autumn migration has brought them over our borders, however, he has cast off many of his fine feathers, together with his gallant manners, and closely resembles the duck in all but character. He is ever a selfish idler, while she attends to all the drudgery of making the nest in the marshy border of the lake; of incubating from six to fourteen pale greenish buff eggs during four weeks of the closest confinement; of caring for the large brood and teaching the ducklings all the family arts.

Shovelers are expert swimmers and divers, though they "tip up" rather than dive for food; they are good walkers also, when we see them in the corn fields, and almost as swift on the wing as a teal. *Took, took; took, took*, that answers as a love song and the expression of whatever passing emotion the ordinarily silent birds may voice, was likened by Nuttall to "a rattle, turned by small jerks in the hand."

Like most other ducks of this subfamily, the shoveler is not common in the northern Atlantic states. Salt water never attracts it; but, on the contrary, it rejoices in lakes, sluggish rivers and streams, isolated grass-grown ponds, and even puddles made by the rain. In the sloughs and lagoons of the lower Mississippi Valley it is still fairly common all winter, however much it is persecuted by the gunners.

"These birds migrate *across* the country to the western plains where they nest," says Chamberlain, "from North Dakota and Manitoba northward, ranging as far as Alaska." In such remote places, where the hand of the law rarely reaches the nefarious pot hunter, he happily finds the ducks in the very prime of toughness.

Pintail

(Dafila acuta)

Called also: SPRIGTAIL; WINTER DUCK

Length—Male, 25 to 30 inches, according to development of tail. Female, 22 inches.

Male—Head and throat rich olive brown, glossed with green and purple; blackish on back of neck; two white lines, beginning at the crown, border the blackish space, and become lost in the white of the breast and under parts. Underneath faintly, the sides more strongly, and the back heavily marked with waving black lines; back darkest; shoulders black; wing coverts brownish gray, the greater ones tipped with reddish brown; speculum or wing patch purplish green; central tail feathers very long and greenish black. Bill and feet slate colored.

Female—Tail shorter, but with central feathers sharply pointed. Upper parts mottled gray and yellowish and dark brown; breast pale yellow brown freckled with dusky; whitish beneath, the sides marked with black and white; only traces of the speculum in green spots on brown area of wing; tail with oblique bars. In nesting-plumage the drake resembles the female except that his wing markings remain unchanged.

Range—North America at large, nesting north of Illinois to the Arctic Ocean; winters from central part of the United States southward to Panama and West Indies.

Season—Chiefly a spring and autumn migrant, or more rarely a winter visitor, in the northern part of the United States; a winter resident in the south.

No one could possibly mistake the long-tailed drake in fall plumage for any other species; but the tyro who would not confound his dusky mate with several other obscure looking ducks, must take note of her lead colored bill and legs, broad, sharply pointed tail feathers, and dusky under wing coverts. The pintails carry themselves with a stately elegance that faintly suggests the coming swan. Their necks, which are unusually long and slender for a duck; their well poised heads and trim, long bodies, unlike the squat figure of some of their kindred; their sharp wings and pointed tails, give them both dignity and grace in the air, on the land, or in the water, for they appear equally at home in the three elements.

But of such charms as they possess they are exceedingly

chary. In the wet prairie lands and grass-grown, shallow waters which they delight in, hunters find these birds the first to take alarm—troublesomely vigilant, noisy chatterers, with a very small bump of curiosity that discourages tolling or decoys; nervous and easily panicstricken. At the first crack of the gun they shoot upward in a confused, struggling mass that gives all too good a chance for a pot shot. If they had learned to scatter themselves in all directions, to dive under water or into the dense sedges when alarmed, as some ducks do, there would be many more pintails alive to-day; but usually they practise none of these protections. There are men living who recall the times, never to return, when ducks resorted literally by the million to the Kankakee and the Calumet regions; and pintails in countless multitudes swelled the hordes that thronged out of the north in the autumn migration. In spite of their enormous fertility, their strong, rapid flight, their swimming and diving powers, their shyness and readiness to take alarm—in spite of the lavish protection that nature has given them, and of their economic value to man—there are great tracts of country where these once abundant game birds have been hunted to extinction.

From the west and the north sportsmen follow the ducks into the lower Mississippi Valley region and our southern seaboard states, where the majority winter. Widgeons and black ducks often associate with them there. The canvasback, the redhead, the black duck, the teals, and the mallard, while counted greater delicacies, by no means attract the exclusive attention of the pot hunter when pintails are in sight. Given a good cook and a young, fat, tender duck, even Macaulay's schoolboy could tell the result.

It is an amusing sight to see a flock of drakes feeding in autumn, when they chiefly live apart by themselves. Tipping the fore part of their bodies downward while, with their long necks distended, they probe the muddy bottoms of the lake for the vegetable matter and low animal forms they feed upon, their long tails stand erect above the surface, like so many bulrushes growing in the water. They seem able to stand on their heads in this fashion indefinitely; a spasmodic working of their feet in the air from time to time testifying only to the difficulty a bird may be having to loosen some much desired root.

From eight to twelve yellowish olive or pale greenish white

PIN-TAIL DUCK.

eggs are laid near the water, but in dry, grassy land, where the mother, who bears all the family cares, forms a slight depression in the soil, under some protecting bush, if may be, and lines it with feathers from her breast.

Wood Duck

(*Aix sponsa*)

Called also: SUMMER DUCK; BRIDAL DUCK; WOOD WIDGEON; TREE DUCK; ACORN DUCK

Length—17 to 19 inches.

Male—Crown of head, elongated crest, and cheeks golden, metallic green, with purple iridescence; a white line from base of bill over the eye, and another behind it, reach to the end of crest; throat, and a band from it up sides of head, white; breast rich reddish chestnut spotted with white; white underneath, shading into yellowish gray on the sides, which are finely marked with waving lines of black; strong black and white markings on long feathers at back of the flanks on the sides. Upper parts dark, iridescent and purplish, greenish brown, a white crescent and a black one in front of wings, which are glossed with purple and green and tipped with white; wing patch purplish blue edged with white; spot at either side of base of tail, chestnut purple. Bill pinkish, red at the base, black underneath and on ridge and tip. Legs yellow.

Female—Smaller. Crest and wing markings more restricted; head dusky with purplish crown; throat, patch around eye, and line backward, white; breast and sides grayish brown, streaked with buff; underneath white; back olive brown glossed with greenish and purple. Young drake resembles the female.

Range—"North America at large, but chiefly in the United States, breeding throughout its range, wintering chiefly in the south." (Coues.)

Season—Summer resident.

This most beautiful of all our ducks, if not of all American birds, in the opinion of many, that Linnæus named the bride (*sponsa*), although it is the groom that is particularly festive in rich apparel and flowing, veil-like crest, confines itself to this continent exclusively; neither has it a counterpart in Europe or Asia as most of our other ducks have. It is an independent little creature with a set way of doing things quite apart, many of them, from family traditions. For instance, it nests in trees rather than on the ground and walks about the limbs like any song bird; it never quacks, but has a musical call all its own; the lovers do not cease to be such after the incubation begins—to name only a few of the wood duck's peculiarities.

Arriving from the south, already mated, in April, a couple prepare to spend the summer with us by selecting a home immediately; an abandoned hole where an owl, a woodpecker, a squirrel, or a blackbird has once nested, answers admirably; or, if such a one be not available, the twigs, grasses, leaves, and feathers that would have lined an excavation are woven into a loose, bulky nest placed among the branches. Deep woods near water, or belted waterways far away from the sea coast, are preferred localities.

How the plump, squat, little mother can work her way in and out through the small entrance to the hole where, for four weary weeks, she sits on from eight to fourteen ivory eggs, is a mystery. It is usually far too narrow for her, one would think, and yet she evidently has no desire to make it larger, as she easily might do by pecking at the soft, decayed wood. The handsome drake on guard in a tree near by calls *peet, peet, o-eek, o-eek* to encourage her or warn her of any threatened danger, to which a faint, musical response, like the pewee's plaint, comes from the hole where she sits brooding. Many endearments pass between the couple, but there is no division of labor, for no self respecting drake would possibly allow his affection to overrule his disinclination for work. The duck attends to all household duties, evidently flattered and content with the vocal expressions of her lord's regard and his standing around and looking handsome, which cost him nothing. The constant moving of his tail from side to side, when perching, is his most energetic effort.

When the fluffy little ducklings finally emerge from the shell, it is the mother who has the task of carrying the numerous brood to water. Often the nest is in a tree overhanging a lake, a quiet stream, or pond, in which case she has only to tumble the babies out of their cradle into the water, where they are instantly at home. But if the tree stands back from the water's edge, one by one she must carry them in her bill and set them afloat, while the father swims around them on guard, proud of them,

no doubt, proud of his energetic busy mate, but doubtless most proud of himself. Wood ducks become exceedingly attached to their home. They return year after year to the same hole to nest, regardless of approaching civilization, the diversion of a water course for factory purposes, the whistle of the locomotive. It is the gunner alone who drives them to a more secluded asylum. On the outskirts of villages these ducks often fearlessly enter the barnyard to pick up the poultry's grain; and there are plenty of instances where they have been successfully domesticated.

In July the drake withdraws to moult his bridal garments, leaving his overworked mate to lead the ducklings about on land and water in quest of seeds of plants, wild oats or rice, roots of aquatic vegetables, acorns, and numerous kinds of insects. The small coleoptera that skips and flies so nimbly along the surface of still inland waters, among the sedges and the lily pads, is ever a favorite morsel, a fact that testifies to the expert swimming of this duck. By September the drake comes out from his exile clad in plumage resembling the duck's, but still more brilliant than hers, and retaining the white throat markings. As the young birds have been gradually shedding their down through the summer and putting on feathers like their mother's, the family likeness in each individual is now most marked. Wood ducks, if ever gregarious, are so in autumn, when flocks begin to assemble early for the southern migration; but at the north we see only family parties preparing for the journeys that are made at twilight and by night, although in the south we hear of companies sometimes numbering a hundred or more. Unhappily, their sweet, tender flesh is in a demand exceeding the legitimate supply in every state they pass through.

"The wood duck is far too beautiful a bird to be killed for food. Its economic value is too small to be worth a moment's consideration," says Mr. Shields. "I would as soon think of killing and eating a Baltimore oriole or a scarlet tanager as a wood duck, and I hope to see the day when the latter will be protected all the year round by the laws of all the states in the Union and of all the provinces of Canada."

SEA AND BAY DUCKS
(Subfamily Fuligulinæ)

Redhead
(Aythya americana)

Called also: AMERICAN POCHARD

Length—19 to 20 inches.

Male—Well rounded head and throat, bright reddish chestnut, with coppery reflection; lower neck, lower back, and fore parts of body above and below, black; rest of the back, sides, and shoulders waved with black and white lines of equal width, that give the parts a silvery gray aspect. Wings brownish gray, minutely dotted with white; wing patch ashy, bordered with black; wing linings chiefly white like the under parts. Bill, which is less than two inches long, dull blue, with a black band at end. Legs and feet grayish brown.

Female—Upper parts dull grayish brown; darker on lower back, the feathers edged with buff or ashy, giving them a mottled appearance; forehead wholly brown; line behind eye and cheeks reddish; upper throat white; neck buff; breast and sides grayish brown washed with buff, and shading into white underneath; an indistinct bluish gray band across end of bill.

Range—North America at large; nesting from California and Minnesota northward, and wintering south of Virginia to West Indies.

Season—Spring and autumn migrant, or winter visitor.

Caterers not up in ornithology very often have this common wild duck of the market stalls palmed off on them, at a fancy price, for canvasbacks; and the tyro on the duck shores of the Chesapeake and our inland lakes just as frequently confuses these two species. Here are a few aids to identification offered in the interest of science, and not because any sympathy need be felt for one who is compelled to eat a redhead, the peer of any table duck.

The bill of the canvasback is a full half inch longer than that of the redhead. The longer, narrower head of the former slants gradually backward from the bill, while that of the latter rises more abruptly, giving the duck a full, round forehead. The plumage on the head and neck of the redhead is decidedly rufous, without any black, whereas the canvasback is rufous brown on those parts, except on the chin and crown, which are blackish. The white lines on the almost white back of the canvasback are wider than those of the redhead, whose black and white waves are of equal width, and look silvery. Usually canvasbacks are larger, heavier birds, but not always. Finally, the females may be distinguished by the difference in their backs, the canvasback duck having wavy white lines across a grayish brown ground, while the redhead is dull mottled brown and buff above. Unscrupulous dealers have a trick of pulling out the telltale feathers, however, which leaves the housekeeper only the shape of the duck's head and bill to guide her choice and protect her purse. As both these species frequent the same bodies of water, constant opportunities for comparisons are offered to that very small minority, alas, who are more interested in the study of the living duck than in the flavor of one roasted.

When the ice begins to form at the far north, where the redheads have spent the summer, great flocks come down to us, eschewing New England with unaccountable perversity, and taking up a temporary residence in the smaller lakes that drain into the Great Lakes and the larger western rivers, before descending to the Chesapeake shores—the duck's paradise—and the lagoons of our southern states, where they pass the winter. It must not be for a moment supposed that because this group of birds is called sea and bay ducks they are found exclusively around salt water. On the contrary, many are more abundant in the interior than along the coast. The classification has reference to the lobe, or web, of these birds' feet, which are most fully equipped for swimming and diving. The redhead and all its immediate kin plunge through deep water. Those that feed in the great beds of wild celery, or vallisneria, gain a peculiar sweetness and delicacy of flesh. In regions where this eel-grass does not grow —as in California, for example—and the redhead must live upon fish, lizards, tadpoles, and the coarser aquatic vegetables, it enjoys no patronage whatever from epicures; whereas in the Mississippi Valley and the Chesapeake, where this "celery" grows most abundantly, gunners shoot thousands on thousands to supply the demand.

A great troop of redheads flying in a close body along the coast in autumn makes a roar like thunder, as their long, strong wings beat the air in unison. Alighting on the waters above their feeding ground, they are at first restless, alert, constantly wheeling about in the air to reconnoitre, before settling down to enjoy themselves with an easy mind. If they have been decoyed to the duck shores at daybreak by gunners screened behind blinds, or tolled within range, a volley welcomes them ; the survivors of the flock quickly outrace sight itself; the wounded escape by diving; and well-trained dogs, plunging through the icy water, bring in to shore the tax that has been levied on the "bunch." Sink boats and reflectors, employed by market shooters who turn sport to slaughter, must soon be suppressed if there is to be any sport left—a doubtful possibility at the present rate of decrease.

In the sloughs and shallow waters of the interior—too shallow for diving—the redheads dabble about like any pond ducks, and tip up one extremity while the other probes the muddy bottom for food. It is in such marshy waters at the north that they build a nest among the rank herbage close to shore. Here it sometimes rests on the water, or else very close beside it; for these ducks are poor walkers, and the mother chooses to glide off the large nestful of buff eggs directly into her natural element. As usual, the drake keeps at a distance when there is any work to be done. Their call note is a sort of hiss, suggesting their ancestors, the reptiles, on the one hand, and their immediate kin, the geese, on the other.

Canvasback
(Aythyra vallisneria.)

Called also: WHITE BACK ; BULL-NECK

Length—21 inches ; generally a little larger than the redhead.

Male—Head and neck dark reddish brown, almost black on crown and chin. A broad band of black encircles breast and upper back; rest of the back and generally wing coverts silvery gray, almost white, the plumage being white, broken up with fine wavy black lines often broken into dots across

CANVASBACK DUCK

the feathers; white underneath; sides dusky; pointed tail feathers darkest slate. Bill, longer than head and shaped like a goose's, from 2.50 to 3 inches in length. Eyes red; feet bluish gray.

Female—Head, neck, collar around upper back and breast, cinnamon or snuff brown; lighter on the throat; back and sides grayish brown marked with waving white lines; white underneath.

Range—North America at large, nesting from the Rocky Mountains and the upper tier of our western states to Alaska and the farthest British possessions, and wintering in the United States, especially in the Chesapeake and middle Texas regions, southward to Central America.

Season—Autumn and spring migrant, and winter resident.

"There is little reason for squealing in barbaric joy over this over-rated and generally underdone bird," says Dr. Coues; "not one person in ten thousand can tell it from any other duck on the table, and only then under the celery circumstances." Yet it is this darling of the epicures that, with the stewed terrapin of Maryland kitchens, has conferred on Baltimore the title of the "gastronomic capital" of our country. There, where it is brought to market fattened on the wild celery in the Chesapeake, it is in its prime a tender, delicately flavored duck, but not one whit more delicious than the canvasbacks taken in Wisconsin, for example, where the celery beds cover hundreds of miles; or the redheads that feed in the same place; or, indeed, than many of the river and pond ducks unknown to the gourmands of Maryland. Redheaded ducks are constantly palmed off at fancy prices by unscrupulous dealers on uninformed caterers, who suffer only in pocket-book by the deception; but the novice who wishes to get what he is paying for is referred to the preceding biography to learn the distinguishing marks of these close associates.

After all it is the food it lives upon, and not its species, that is responsible for any duck's flavor. Canvasbacks have an immense range, and where no wild celery grows, and they must harden their muscles in the active pursuit of fish, lizards, and other animal diet, they become as tough and rank as a merganser, ignored and even despised members of the duck clan these *précieuses ridicules*.

The wild celery, or *vallisneria spiralis*, which is no celery at all, but an eel grass growing entirely beneath the water, took its name from Antonio Vallisneri, an Italian naturalist, and it was passed on as a specific name to the canvasback. When fattened upon it a brace of these ducks often weigh twelve pounds. To secure its buds and roots, the only parts they eat, they must dive and remain a long time under water, only to be robbed on their return many times by the bold baldpates that snatch the celery from their bills the instant their heads appear above water. Several duck farms have been recently established where the common plebeian domestic duck is fed on celery and fattened for the market. Then this vulgar bird is served up at hotels and restaurants as canvasback, at from three to five dollars a plate, and no one, not even the epicure, can tell the difference.

Exceedingly shy, wary, restless scouts, the canvasbacks are decoyed within gun range only by the sportsman's subtlest wiles. It is no part of the plan of this book to assist in the already rapid extermination of our game birds by detailing the manifold schemes devised for their capture, which when fully investigated vastly increase our respect for a bird that can save its neck in passing through this land of liberty. This and other diving ducks that wear thick feathered chest protectors may fall to the water, stunned by the sportsman's shot, but quickly revive, and escape under water; while the retriever, nonplussed by their disappearance, is blamed for his stupidity.

One would imagine our ornithologists were writing cookbooks, to read their accounts of this duck whose habits have been little studied beyond its feeding grounds in the United States. Its life history is still incomplete, although its nesting habits are supposed to be identical with those of the redhead, and its buff eggs are known to have a bluish tinge. It is in death that the canvasback is glorified.

Greater Scaup Duck

(Aythyra marila nearctica)

Called also: AMERICAN SCAUP; BROADBILL; BLACKHEAD; BLUEBILL; RAFT DUCK; FLOCKING FOWL; SHUFFLER.

Length—17.50 to 20 inches.

Male—Black on upper parts, with greenish and purplish reflections on head; lower back and about shoulders waved with black and white; under parts white, with black waving bars on sides of body and near the tail; speculum, or wing mirror, white. Bill dull blue, broad, and heavy; dark, slate-colored feet.

Female—A white space around base of bill, but other fore parts rusty, the rusty feathers edged with buff on the breast; back and shoulders dusky, and the sides dark grayish brown, finely marked with waving white lines; under parts and speculum white.

Range—North America at large; nesting inland, chiefly from Manitoba northward; winters from Long Island to South America.

Season—Common spring and autumn migrant, and winter resident south of New England and the Great Lakes.

If the number of popular names that get attached to a bird is an indication of man's intimacy with it, then the American scaup is among the most familiar game birds on the continent. It is still a mooted question whether the word scaup refers to the broken shell fish which this duck feeds upon when wild celery, insects, and fry are not accessible, or to the harsh, discordant *scaup* it utters, but which most people think sounds more like *quauck*. Its broad, bluish bill, its glossy black head, its not unique habit of living in large flocks, its readiness to dive under a raft rather than swim around one, and its awkward, shuffling gait on land, where it rarely ventures, make up the sum of its eccentricities set forth in its nicknames.

Gunners in the west and on the Atlantic shores from Long Island southward, especially in the Chesapeake, where wild celery abounds, find the bluebills among the most inveterate divers: they plunge for food or to escape danger, loon fashion, and when wounded have been known to cling to a rock or tuft of sedges under water with an agonized grip that even death did not unfasten. They do not rise with ease from the surface of the water, which doubtless often makes diving a safer resort than flight. Audubon spoke of their "laborious flight;" but when once fairly launched in the air, their wings set in rigid curves, they rush through the sky with a hissing sound and a rate of speed that no amateur marksman ever estimates correctly. They are high flyers, these bluebills; and as they come swiftly winding downward to rest upon the bays of the seacoast or large bodies of inland waters, they seem to drop from the very clouds.

Sea and Bay Ducks

No dabblers in mud puddles are they: they must have water deep enough for diving and cold enough to be exhilarating. Diving ducks feed by daylight chiefly, or they would never be able to distinguish a crab claw from a celery blade; but they also take advantage of moonlight for extra late suppers. In the Chesapeake region flocks of ducks that have "bedded" for the night rise with the rising moon, and disport themselves above and below the silvery waters with greater abandon even than by day. Owing to the thick feathered armor these ducks wear, the sportsman often counts birds shot that, being only stunned, are able to escape under water.

It is only when the nesting season has closed that we find the bluebills near the seacoast. They build the usual rude, duck-like cradle—or, rather, the duck builds it, for the drake gives nursery duties no thought whatever—in the sedges near an inland lake or stream, where this ideal mother closely confines herself for four weeks on from six to ten pale olive buff eggs. Nuttall observed that "both male and female make a similar grunting noise" (the *quauck* or *scaup* referred to), "and have the same singular toss of the head with an opening of the bill when sporting on the water in spring."

. . . .

The Lesser Scaup Duck *(Aythya affinis)*, Creek Broadbill, Little Bluebill, and so on through diminutives of all the greater scaup's popular names, may scarcely be distinguished from its larger counterpart, except when close enough for its smaller size (sixteen inches), the purplish reflections on its head and neck, and the heavier black and white markings on its flanks to be noted. Apparently there is no great difference in the habits of these frequently confused allies, except the preference for fresh water and inland creeks shown by the lesser scaup, which is not common in the salt waters near the sea at the north, and its more southern distribution in winter. Chapman says: "It is by far the most abundant duck in Florida waters at that season, where it occurs in enormous flocks in the rivers and bays along the coasts."

. . . .

The Ring-necked Duck *(Aythya collaris)*, or Ring-necked Blackhead, Marsh Bluebill, Ring-billed Blackhead, and Bastard Broadbill, as it is variously called, though of the same size as the lesser scaup, may be distinguished from either of its allies by a broad reddish brown collar, a white chin, entirely black shoulders, gray speculum on wings, and a bluish gray band across the end of the broad, black bill, which are its distinguishing marks. While the female closely resembles the female redhead, its smaller size, darker brown coloration, gray speculum, indistinct collar, and the shape and marking of its bill, are always diagnostic with a bird in the hand. This broadbill is almost exclusively a fresh water duck: not an abundant bird anywhere, apparently, even in the well-watered interior of this country and Canada, which is all ducks' paradise; and mention of its occurrences are so rare along the Atlantic coast as to make those seem accidental. On the fresh water lakes of some of the southern Atlantic states it is as abundant in winter, perhaps, as it is anywhere. Its classification among the sea and bay ducks has reference only to the full development of its feet.

It was Charles Bonaparte, Prince of Canino, who first named this duck, which had been previously confounded with the two other broadbills, as a distinct species; and we are still indebted to that tireless enthusiast for the greater part of our information concerning it, which is little enough. So far as studied, its habits differ little from those of its allies. At the base of the head, a few long feathers, scarcely to be distinguished as a crest, are constantly erected as the bird swims about on the lake with its neck curved swan fashion; and Audubon tells of its "emitting a note resembling the sound produced by a person blowing through a tube." Like many another duck, there is more interest shown in this one's flavor than in its life history.

American Golden-eye
(*Glaucionetta clangula americana*)

Called also: WHISTLER; WHISTLE WING; BRASS-EYED WHISTLER; GREAT HEAD; GARROT.

Length—17 to 20 inches.
Male—Head and short throat dark, glossy green; feathers on the former, puffy; a round white space at base of bill; neck all around, breast, greater part of wings, including speculum and under parts, white; wing linings dusky; rest of plumage

Sea and Bay Ducks

black. Feet orange, with dusky webs; bill black or blackish green, and with large nostrils; iris bright golden.

Female—Much smaller; head and throat snuff color, and lacking the white space near the bill; fore neck white; upper parts brownish black; under parts white, shading into gray on sides and upper breast, which are waved with gray or brown; speculum white, but with less white elsewhere on wings than male's. Bills variable.

Range—North America, nesting from our northern boundaries to the far north, and wintering in the United States southward to Cuba.

Season—Winter resident, also spring and summer migrant in United States.

The Indians of Fraser valley tell a story of two men in one of their tribes who began to discuss whether the whistling noise made by this duck was produced by its wings or by the air rushing through its nostrils. The discussion waxed warm and furious, and soon others joined in. Sides were taken, one side claiming that the drakes, with their larger nostrils, make a louder noise than their mates, and that the scoters, which also have large nostrils, make a similar whistling sound when flying. The other side contended that whereas the wings of all ducks whistle more or less, the incessant beating of the golden-eye's short, stiff wings, that cut the air like a knife, would account for the louder music. Before long the entire crowd became involved in the dispute; tomahawks were brandished and a free fight followed, according to Allan Brooks, in which a majority of the warriors were killed without settling the question—an excellent story for the Peace Societies.

Pale Faces, backed by scientific investigation, take sides with the wing whistler party. The golden-eye, in spite of its short, heavy body and small wings, covers immense distances, ninety miles an hour being the speed Audubon credited it with, and a half mile the distance at which he distinctly heard the whistle. Although the drake, at least, has every requisite in his vocal organs for making a noise, and the specific name, *clangula*, entitles him to a voice, it has never been lifted in our presence. But then this duck has been very little studied in its nesting grounds, where, if ever, a bird gives utterance to any pent-up emotion. In the desolate fur countries at the far north of Europe and America, the golden-eye duck makes a nest in a stump or hollow

AMERICAN GOLDENEYE.

tree, close by the lake or river side, and covers over her large clutch of pale bluish eggs with down from her breast. As usual in the duck tribe, the drake avoids all nursery duties by joining a club of males that disport themselves at leisure during the summer moult.

Wonderfully expert swimmers and divers, their fully webbed feet, that make these accomplishments possible, so interfere with their progress on land that they visit it only rarely. One can distinctly hear the broad webs slap the ground, as, with wings partly distended to help keep a balance, the golden-eye labors awkwardly on by jerks to reach the water, where not even the loon is more at home. As the golden-eye's flesh is rank and fishy and tough, owing to the small proportion of vegetable food it eats, and the large amount of exercise it must take to secure active prey, there can be no excuse for the sportsman's hunting it; and, happily, there is apt to be scant reward for his efforts.

Exceedingly shy and wary, with a sentinel on the constant lookout, and associated only with those ducks that are as quick to take alarm as themselves, the whistlers are among the most difficult birds to approach. They dive at the slightest fear, swim under water like a fish, or, bounding upward with a few labored strokes from the surface of the lake, make off at a speed and at a height the tyro need not hope to overtake with a shot. During the late autumn migration the males precede their discarded mates and young by a fortnight. They continue abundant around many parts of our country, inland and on the coast, and enliven the winter desolation after most other birds have deserted us for warmer climes.

Barrow's Golden-eye (*Glaucionetta islandica*), a more northern species, that is often seen in the west, may scarcely be told from the common whistler either in features or habits. A crescent-shaped white spot at the base of the bill of the drake and more purplish iridescence on his head are his distinguishing marks; but the small females of these two species are believed to be identical. In the region of the salmon canneries these ducks lose some of their native shyness and boldly gorge themselves on the decaying fish. Allan Brooks writes that "the note is a hoarse croak." Doubtless the common golden-eye makes some such noise also, or that close student, Charles

Bonaparte, would never have named it *clangula*. "They have also a peculiar mewling cry," Brooks adds, "made only by the males in the mating season."

Bufflehead

(Charitonetta albeola)

Called also: BUTTER-BALL; BUTTER-BOX; SPIRIT DUCK; LITTLE DIPPER; BUFFALO-HEADED DUCK.

Length—13.50 to 15 inches.

Male—A broad white band running from eye to eye around the nape of neck; rest of head with puffy feathers, and, like those on throat, beautifully glossed with purple, blue, and green iridescence. Other upper parts black; neck all around, wings chiefly, and under parts wholly, white. Bill dull blue; feet flesh color.

Female—Blackish brown above, with white streak on each side of head; whitish below. Smaller than male.

Range—North America at large, nesting from Dakota, Iowa, and Maine northward to the fur countries; winters from the southern limit of its nesting range, or near it, to Mexico and the West Indies.

Season—Transient spring and autumn visitor, or winter resident from November to April.

Not even a grebe or loon is more expert at diving "like a flash" than this handsome little duck. Samuels says that "when several of these birds are together one always remains on the surface while the others are below in search of food, and if alarmed it utters a short *quack*, when the others rise to the surface, and on ascertaining the cause of alarm all dive and swim off rapidly to the distance of several hundred feet."

A bufflehead overtakes and eats little fish under water or equally nimble insects on the surface, probes the muddy bottom of the lake for small shell fish, nibbles the sea-wrack and other vegetable growth of the salt-water inlets, all the while toughening its flesh by constant exercise and making it rank by a fishy diet, until none but the hungriest of sportsmen care to bag it. Yet this duck is more than commonly suspicious and shy. It will remain just below the surface, with only its nostrils exposed to the air, for an hour after a severe fright, rather than expose its fat little body, that it prizes more highly than do those who know its worth. In any case a shot is more likely to stun than to kill a bufflehead, that, like most other diving birds, is armored with a thick, well-nigh impenetrable suit of feathers. It may fall as if mortally wounded, but the cold water usually revives it at once, and the expectant gunner looks for his victim many yards from where it is safely recovering from its recent excitement.

Because it can so illy protect itself on land, for it is a wretched walker, and doubtless also because it chooses to nest in countries where the fox and other appreciative eaters of its flesh abound, the bufflehead enters a hollow tree to lay her light buff or olive eggs. Here she sits, often in the dark, for four weary weeks, quite ignored by the mate that in February almost bobbed his head off in his frantic efforts to woo her. It is she that must carry the large brood of ducklings in her bill to the water, teach them all she knows on it, and count herself well rewarded if her plumpest babies do not fall into the jaws of a pike ready to swallow the little divers, but are spared to migrate with her to open waters when the ice locks up their food at the north.

Old Squaw

(Clangula hyemalis)

Called also: OLD WIFE; SOUTH-SOUTHERLY; LONG-TAILED DUCK; OLD INJUN; SCOLDER; OLD MOLLY; OLD BILLY; COCKAWEE

Length—Variable, according to development of tail—18 to 23 inches.

Male—*In winter:* Blackish on back, breast, and tail, whose four middle feathers are long and narrow; sides of the head grayish brown; rest of head, neck all around, upper back, shoulders, and underneath, white; no speculum on grayish wings. Bill with large orange-colored patch; feet dusky blue; *In summer:* Sides of head white; top of head, throat, breast above and below, back and shoulders, black; white underneath. Tail longer than in winter.

Female—No elongated feathers in tail, which consists of fourteen feathers coming to a point; head, neck, and upper parts, dusky brown, with grayish patch around the eye and one on side of neck; breast grayish, shading to white below;

the feathers on the upper parts more or less edged with buff in summer.

Range—"In North America, south to the Potomac and the Ohio (more rarely to Florida and Texas) and California; breeds northward."—A. O. U.

Season—Common winter resident in northern United States; November to April.

Like a crowd of gossiping old women these ducks gabble and scold among themselves all the year round, for in winter, when most voices are hushed, they are the noisiest birds that visit us. In summer, they nest so far north that none but Arctic travellers may hope to study them. Mr. George Clarke, of the Peary expedition, writes of "the old squaw's clanging call" ringing out from the drifting ice cakes where the drakes glided about at no great distance from their brooding mates. *South, south, southerly*, is the cry some people with more lively imaginations than accuracy of ear have heard; but the Indians were nearer right when they "called down" this high flyer with a *hah-ha-way*, part of the full cry written by Mr. Mackay as *o-onc-o-onc-ough, egh-ough-egh*. The other part is not very different from the *honk* of a goose. Most of the duck's popular names, as well as its scientific one, allude to its noisy, talkative habit. At evening, and toward spring when the choice of mates involves great discussion and quarrelling, they make more noise than perhaps all our other sea fowl combined.

The plumage of this duck varies so much with age, season, and sex, that it is well we have some pronounced characteristics to help us in naming our bird correctly. The long tail feathers of the drake are its most striking feature; but the obscure-looking duck has little to distinguish her from the female harlequin, except her white abdomen, which is usually concealed under water.

When migrating from the icy regions that they haunt after all other ducks have left for the south, the old squaws proceed by degrees no faster than Jack Frost compels; so that in season as in plumage they are apt to be exceedingly variable, an open winter keeping them north until late, and a cold autumn driving them from the ice-bound waters to seek their fish, mollusks, and water wrack in the open channels of our larger lakes and rivers and the inlets of the sea. Maritime ducks these certainly are by preference; famous divers and swimmers; strong, swift flyers; noisy, restless, lively fellows, that live in a state of happy commotion; gregarious at all seasons, and strongly in evidence wherever they find their way.

There can be no excuse for killing these fish eaters for their flesh, which is rank and apparently in the very prime of toughness throughout their stay here; but they are clothed with particularly thick, fine, lively feathers that are in great demand for pillows. These form an almost invulnerable armor one would think, yet great quantities of old squaws' down and feathers are bought by upholsterers every year. At the north the mother herself pulls out some of her feathers to cover her pale bluish eggs, concealed in a rude nest in grasses or under some low bush near the shore. When wounded, as the duck flies low and very swiftly along the water, it instantly dives from the wing, according to Mr. Mackay. He tells of seeing many of them towering, "usually in the afternoon, collecting in mild weather in large flocks if undisturbed, and going up in circles so high as to be scarcely discernible, often coming down with a rush and great velocity, a portion of the flock scattering and coming down in a zig-zag course similar to the scoters when whistled down."

. . . .

The Harlequin Duck *(Histrionicus histrionicus)*, also called Lords and Ladies, comes down to our more northern coasts of sea and large inland lakes only when ice has closed its feeding grounds at the north; but no clanging call invites our attention when these gay masqueraders appear on the scene, tricked out in black, white, blue, and reddish brown applied in stripes and spots; and as they keep well out from shore to hunt in our open waters, few get a good look at their fantastic coats before they return to the north to nest. The female can scarcely be distinguished from the female old squaw, except by her dusky under parts. A harlequin's flesh is dark and unpalatable, for fishy food is its staple, and no one not hard pressed by hunger would care to eat it. From the characteristics of habit that distinguish all ducks of this subfamily, the harlequin differs little, except in living near rushing, dashing streams of the Rocky and Sierra Nevada mountains and northward during the nesting season. Six or more yellowish or greenish buff eggs are laid in hollow stumps near the water; and the fact that the young ducklings

Sea and Bay Ducks

are not swept away by the swift current of the stream they take to and live on, without returning to the nest once it is left, testifies to the remarkable propelling power of their feet. These ducks are most expert divers, too, and when alarmed will plunge like a grebe, and swim under water to parts unknown.

American Eider
(Somateria dresseri)
Called also: SEA DUCK

Length—23 inches.

Male—Upper parts white, except the crown of head, which is black, with a greenish white line running into it from behind and a greenish tinge on the feathers at sides of back of head. Upper breast white with a reddish blush; lower breast and all under parts, including tail above and below, black.

Female—Upper parts buffy brown, streaked and varied with darker brown and black; back darkest; breast yellow buff, barred with black, and shading into grayish brown, indistinctly margined with buff underneath.

Range—Nests around Nova Scotia and Labrador, migrating southward in winter to New England and the Great Lakes, more rarely south to Delaware.

Season—Winter visitor.

When resting under our down coverlets on a winter night, or tucked about with pillows on the divan of a modern drawing-room, how many of us give a thought to the duck that has been robbed of her soft warm feathers for our comfort, or take the trouble to make her acquaintance when she brings the brood that were despoiled of their bedding to furnish ours to visit our coast in winter? It may be said in extenuation of our apparent indifference that eiders keep well out at sea, and come at a season when boating ceases to be a pleasure. Then, too, there is little to interest one during the winter in a bird whose chief concern appears to be deep diving. It is on the constant errand of getting mussels and other fish food which the saddle-back gull often snatches from it at the end of an unequal race if the duck does not end it suddenly by plunging under water. It is to Labrador and the north Atlantic islands that one must go to know this bird at home, and most of us are willing to do such travelling in the easy chairs of our library.

Before these ducks have left our shores in March, courting has already begun; sharp contests occur, and the vanquished or superannuated males wander about in milder climates than the mated lovers fly to. Though no drake may be credited with great depth of feeling for his mate, the eider goes to the extreme of helping her make a nest of moss and seaweed among the rocks or low bushes under stunted fir trees, and will even pluck the down from his own breast to cover the eggs when hers has been persistently robbed. *Ha-ho, ha-ho,* he half moans, half coos, in a lackadaisical tone to the busy housewife who replies with a matter-of-fact *quack*, like any prosaic barnyard duck. Until the last one of her bluish or olive gray eggs is laid, the mother plucks no down from her breast; but she will continue to lay, and to cover the new eggs with her feathers, several times over if her nest is robbed, until her poor breast is naked and the drake's down is called into requisition. According to Saunders the average yield of down from a nest in Iceland, where the birds are encouraged and protected by law, is about one-sixth of a pound. The gathering of these live feathers, as they are called, for no one thinks of killing this valuable bird or its allies to take their down which loses its elasticity after death, is an important industry in the northern countries of Europe; but the industry is neglected and unintelligently managed on this side of the Atlantic. When all the eggs and down are taken from a nest repeatedly, the despairing birds abandon it for more remote parts, and never return; whereas hope eternally springs in a breast even where feathers do not, if an egg or two are left the mother. Audubon found large colonies of the American eider nesting in Labrador in April, and gathered some fresh eggs for food in May, when ice was still thick in the rivers. He found both ravens and the larger gulls prowling about the coast ready to suck the eggs and carry off the ducklings before they had mastered the art of diving out of harm's reach.

While the females sit upon their nests the drakes withdraw for a thorough moult, which leaves them so bare of feathers in July that they are sometimes unable to fly. Henceforth they live apart, he in flocks of males, she with small companies of mothers with their broods, which latter are usually the flocks that visit

Sea and Bay Ducks

us in winter, for the hardy old drakes do not often migrate so far south. By August ice has begun to form over their northern fishing grounds, and the flocks move a degree nearer us, flying swiftly and powerfully in a direct course, not far above the water, and almost never over land.

American Scoter
(Oidemia americana)

Called also: BLACK, OR SEA COOT; BOOBY; BLACK SCOTER; BUTTER-BILLED COOT; BROAD-BILLED COOT.

Length—19 to 20 inches.

Male—Entire plumage black, more glossy above. Upper half of bill, which is tumid, or bulging, is yellow or orange at the base.

Female—Sooty brown above, waved with obscure dusky lines; throat and sides of head whitish; dirty white underneath; bill dark, but not bulging nor parti-colored. Young resemble the mother.

Range—Seacoasts and large bodies of inland waters of northern North America; nesting from Labrador inland, and migrating in winter to New England and the Middle Atlantic States and to California.

Season—Winter resident and visitor.

    The three species of coots, or scoters, that come out of the north to visit us in winter have neither fine feathers nor edible flesh to recommend them to popular notice; nor do they seem to possess any unique traits of character or singular habits to excite our lively interest. Their chief concern in life appears to be diving for mussels, clams, small fry, and mollusks in the estuaries of rivers and shallow sounds along our coasts. Some go to large bodies of inland waters for the same purpose. As this active exercise toughens their muscles to a leather-like quality, and as the fish food gives their reddish, dark flesh a rank flavor, the poultry dealer who sells one of these birds to an uninitiated housekeeper for black duck loses a customer

    Most friendly with its own kin, the American coot may usually be found in flocks of white-winged and surf scoters, eiders, and other sea ducks, where they congregate above beds of shell fish; and, at least while in the United States, the habits of all these birds appear to be identical. But they are as shy of men as if their breasts were covered with more desirable meat, and dive when approached rather than take to wing and expose their precious ugliness to an unoffending field-glass. Human friendship is discouraged by them, however much their long list of common names, which are as often applied to one species as another, falsely testifies to their popularity.

Ridgway describes their nests as on the ground, near water, and containing from six to ten pale dull buff or pale brownish buff eggs.

. . . .

The White-winged Scoter or Coot (*Oidemia deglandi*), which is sometimes called Velvet Duck, differs from the preceding in plumage only, in having a white patch under the eye, a white mirror, or speculum, on wings, and orange-colored legs, much the same shade as its protuberant bill, which is feathered beyond the corners of the mouth. Possibly it goes farther away from water than the other scoters to place its nest under a bush on the ground, but the habits of all three species appear to be generally the same, and like those of nearly all sea ducks.

. . . .

The Surf Scoter, or Sea Coot (*Oidemia perspicillata*), has a square white mark on the crown of its head and a triangular one on the nape, to distinguish it from its sombre and rather uninteresting relatives.

Ruddy Duck.
(Erismatura rubida)

Called also—SPINE-TAILED DUCK; SALT WATER TEAL; DUN BIRD

Length—15 to 17 inches.

Male—*In summer:* Crown of head and nape glossy black, chin and sides of head dull white; neck all around and upper parts and sides of body rich reddish brown; lower parts white, with dusky bars; wing coverts, quills, and stiff-pointed tail feathers darkest brown; head small; neck thick. Bill, which is as long as head, broader at tip; wings very

Sea and Bay Ducks

short, and without speculum. In winter the drake resembles female.

Female—Upper parts dusky grayish brown, the feathers rippled with buff ; crown and nape more reddish, and streaked with black ; sides of head and chin white ; throat gray ; under parts white. Young resemble mother.

Range—North America at large ; nesting chiefly north of the United States, but also locally within its range ; winters in the United States.

Season—Spring and autumn migrant ; also locally a winter resident.

The heavy moult this drake undergoes after he deserts his brooding mate transforms him into an obscure, commonplace-looking bird from the faultlessly attired gallant of his courting days ; so that when the ruddy ducks appear on our inland lakes or the estuaries of rivers, shallow bays, and ponds near the sea, there is a close family resemblance between both the parents and the young, none of whom seem worthy bearers of their popular name. But however inconspicuous the feathers, this duck may always be named by its stiff tail quills, that no other bird but a cormorant can match. This curious tail, which is used as a rudder under water, or a vertical paddle, is carried cocked up at right angles to the body when the duck floats about on the surface.

Owing to the ruddy duck's short wings, it is less willing to trust its safety to them when alarmed than most ducks are, and it will quietly dive in grebe fashion, and drop to safe depths before swimming out of range, rather than depend upon the awkward rising from the surface, that must be struggled through before it is safely launched in steady though labored flight along the water. Heading against the wind, it at first seems to run along the surface with the help of rapid wing beats, before it is able to clear the water ; but once fairly started, it flies good distances and at a fair speed. In figure it more closely resembles a plump, squat teal than an ordinary sea duck. The head is so small that the skin of the neck can be easily drawn over it.

Tall sedges near the water's edge make the ideal nesting or hunting resort of these ducks, that feed chiefly on eel grass and other vegetable matter growing either above or below the water in shallow bays and inlets, salt or fresh. It is their habit to drop into these grasses when surprised, and to hide among them, which is one reason why they are supposed to be rare ; whereas they are fairly abundant, though often unsuspected. Numbers of them find their way into large city markets every winter; and especially in the Chesapeake region, or where wild celery abounds, their flesh is tender and well flavored. Happily the species is very **prolific**. Some authorities mention finding as many as twenty yellowish white, rough eggs in the rude nests built by the marshy **lake** or river side ; but ten are a good-sized clutch.

GEESE

(Subfamily *Anserinæ*)

American White-fronted Goose

(*Anser albifrons gambeli*)

Called also: LAUGHING GOOSE; SPECKLE-BELLY; GRAY BRANT; PRAIRIE BRANT

Length—27 to 30 inches.

Male and Female—Upper part and fore neck brownish gray, the edgings of the feathers lighter; a white band along forehead and base of bill bordered behind by blackish; lower back, nearest the tail, almost white; wings and tail dusky; sides like the back; breast paler than throat, and marked, like the white under parts, with black blotches; bill pink or pale red; feet yellow; eyes brown. Immature birds, which are darker and browner than adults, lack white on forehead and tail coverts, also the black patches on the under parts.

Range—North America; rare on Atlantic coast; common on the Pacific slope and in the interior; nesting in the far north, and wintering in the United States southward to Mexico and Cuba.

Season—Spring and autumn migrant or winter resident on the plains and westward to the Pacific.

A long, clanging cackle, *wah, wah, wah, wah,* rapidly repeated, rings out of the late autumn sky, and looking up, we see a long, orderly line of laughing geese that have been feeding since daybreak in the stubble of harvested grain fields, heading a direct course for the open water of some lake. With heads thrust far forward, these flying projectiles go through space with enviable ease of motion. Because they are large and fly high, they appear to move slowly; whereas the truth is that all geese, when once fairly launched, fly rapidly, which becomes evident enough when they whiz by us at close range. It is only when rising against the wind and making a start that their flight is actually slow and difficult. When migrating, they often trail across the clouds like dots, so high do they go—sometimes a thousand feet or more, it is said—as if they spurned the earth. But as a matter of fact they spend a great part of their lives on land; far more than any of the ducks.

On reaching a point above the water when returning from the feeding grounds, the long defile closes up into a mass. The geese now break ranks, and each for itself goes wheeling about, cackling constantly, as they sail on stiff, set wings; or, diving, tumbling, turning somersaults downward, and catching themselves before they strike the water, form an orderly array again, and fly silently, close along the surface quite a distance before finally settling down upon it softly to rest.

Such a performance must be gone through twice a day, once after their breakfast, begun at daybreak, and again in the late afternoon, on their return from their inland excursion, which may be to stubble fields, or to low, wet, timbered country, or to bushy prairie lands. Not only the farmer's cereals, but any sort of wild grain and grasses, berries, and leaf buds of bushes, these hearty vegetarians nip off with relish. When we see them on shallow waters, with tail pointing skyward and head and neck immersed, they are probing the bottom for roots of water plants, particularly for a sort of eel-grass that they fatten on, or for gravel, and are not eating mollusks or any sort of animal food, as is sometimes said.

But fatal consequences await on ducks and geese alike that do not know enough to toughen their flesh and make it rank by a fish diet. White-fronted geese, delicious game birds of the first order, were once abundant during the migrations in the Chesapeake country, where they freely associated with the snow goose and the Canada species, just as they do in the far west to-day; but the sportsman must now travel to the Great Lakes or the plains, or, better still, to California, their favorite winter resort, if he would see a good sized flight above the stubble fields, in which, hidden in a hole, and with flat decoys standing all about him, he waits, cramped and breathless, for the cackling flock to come within range.

The stupidity of this bird is more proverbial than real. If any one doubts this, let him try to stalk one when it is feeding in the fields, or listen to the tales of woe the California farmers tell of its provoking vigilance and cleverness.

Geese

Snow Goose

(Chen hyperborea nivalis)

Called also: WHITE BRANT; WAVEY; BLUE-WINGED GOOSE

Length—27 to 35 inches.
Male and Female—Entire plumage white, except the ends of wings, which are blackish, and the wing coverts, which are grayish; bill carmine; legs dull red. Immature birds have feathers of upper parts grayish with white edges.
Range—North America at large, nesting in the far north (exact sites unknown), and migrating to the United States to pass the winter. More abundant in the interior and on the Pacific slope than on the Atlantic, north of Virginia.
Season—Spring and autumn migrant, April and October; or winter resident in milder parts of the United States to Cuba.

The dullest imagination cannot but be quickened at the sight of a great flock of these magnificent birds streaming across the blue of an October sky like a trail of fleecy white clouds. Such a sight is rare indeed to people on the Atlantic coast north of the Chesapeake; but in the Mississippi valley during the migrations, on the great plains, and in parts of California all winter, fields are whitened by them as by a sudden fall of snow. Lakes in Minnesota may still be seen reflecting their glistening whiteness as if snow peaks were mirrored there; and in the Sacramento and San Joaquin valleys, in Oregon and beyond, they are still sufficiently abundant to be hunted on horseback by the indignant farmers, who see no beauty in their plumage to compensate them for their devasted fields of winter wheat that the hungry flocks nip off close to the ground. But like most other choice game birds, the snow goose is fast disappearing. Who that knows how rapid this decrease is ever expects to see such flocks of these superb fowl as gladdened the eyes of Lewis and Clarke when they reached the mouth of the Oregon?

Closely associated with the white-fronted and the Canada geese, the white brant may be named, even when too high up in the sky at the twilight of dawn or evening for us to see its darktipped wings and white plumage, by the higher pitched, noisier cackling that distinguishes its voice from that of the laughing goose and the mellow *honk* of the Canada brant. It migrates by night and day; observes punctual meal hours like the the rest of its kin; keeps a sentinel always on guard while it feeds in the grain fields or roots among the rushes on the tide-water flats and grassy patches bordering streams; circles, gyrates, tumbles, and floats above the water on returning from its feeding grounds. In short, it behaves quite as other geese do when intoxicated with food.

While it is supposed the white brant nests somewhere in the region of the Barren Grounds between the Mackenzie basin and Greenland, the nest and eggs are still unknown in that little-visited country beyond the north wind (*hyperboreus*), as the bird's name indicates.

. . . .

The Lesser Snow Goose (*Chen hyperborea*), a smaller species, identical in plumage with the preceding, and very like it in habits, nests in Alaska, and wanders down the Pacific coast in winter, eastward to the Mississippi and southward to the Gulf.

Canada Goose
(Branta canadensis)

Called also:—WILD GOOSE; GRAY GOOSE; *HONKER*

Length—From 1 yard to 43 inches.

Male and Female—Head and neck black, a broad white band running from eye to eye under the head; mantle over back and wings grayish brown, the edges of feathers lightest; breast gray, fading to soiled white underneath. Female paler; tail, bill, and feet black.

Range—North America at large, nests in northern parts of the United States and in the British possessions; winters southward to Mexico.

Season—Chiefly a spring and autumn migrant, north of Washington; although a few remain so late (December) and return so early (March) they may almost be said to be winter residents north as well as in the south. The most abundant and widely distributed of all our wild geese.

Heralded by a mellow *honk, honk,* from the leader of a flying wedge, on come the long-necked wild geese from their northern nesting grounds, and stream across the sky so far above us that

Geese

their large bodies appear like two lines of dark dots describing the letter V. In spite of their height, which never seems as great as it actually is because of the goose's large size, one can distinctly hear the *honk* of the temporary captain—some heavy veteran—answered in clearer, deeper tones, as the birds pass above, by the rear guardsmen in the long array that moves with impressive unison across the clouds. Often the fanning of their wings is distinctly audible too. The migration of all birds can but excite wonder and stir the imagination; but that of the wild goose embarked on a pilgrimage of several thousand miles, made often at night, but chiefly by broad daylight, attracts perhaps the most attention. Sometimes the two diverging lines come together into one, and a serpent seems to crawl with snake-like undulations across the sky; or, again, the flock in Indian file shoots straight as an arrow. It is as a bird of passage that one thinks of the goose, however well one knows that it remains resident in many places at least a part of the winter.

A slow drift down a slope of a mile or more, on almost motionless wings, brings them to the surface with majestic grace, and flying low until the precise spot is reached where they wish to rest, they settle on the water with a heavy splash. Usually they stop flying near sunset to feed on the eel-grass, sedges, roots of aquatic plants, insects, and occasionally on small fish, or on the wheat, corn, and other grain that has dropped among the stubble in the farmer's fields, and the berries, grass, and leaf buds they find in swamps and bushy pastures. Quantities of gravel are swallowed with their food. After a good supper they return to the water, preferably to a good-sized lake, to sleep, and there they float about with head tucked under wing until daybreak, when another flight must be made inland to secure a breakfast. These two regular daily flights are characteristic of all the geese.

Such punctuality at meals is confidently reckoned upon by the sportsman, who is thereby saved unnecessary waiting as he crouches, cramped and cold, in a pit among the stubble and concealed by a blind. These holes are about thirty inches in diameter and about forty inches in depth. There are no birds with keener, more suspicious eyes; no sentinel of a flock more on the alert, unless it be the sandhill crane, that often feeds with them and is their ally; no game birds more wary when the sportsman tries to stalk them than these; and so no one can possibly

CANADA GOOSE.
½ Life-size.

appreciate the expression "a wild goose chase" who has not hunted them. The goose is by no means the dolt tradition says it is. The ordinary methods of hunting water-fowl do not answer with it, and in different parts of the country a different ruse is practiced to secure its flesh. Strangely enough, ducks and geese alike, that are thrown into a state of panic at sight of a man or dog, show no fear whatever of cows; and taking advantage of this fact, gunners often hide behind cattle, or lead a horse or an ox to get within range. On the great plains and in California, oxen trained for the purpose screen the hunters on horseback, and walk straight into the flocks of Canada, snow, and laughing geese that have been lured by live or artificial decoys placed in some good feeding ground. Geese are not only gregarious, but extremely sociable to their kin and to other birds as quick to take alarm as they. A constant gabbling goose-talk is overheard wherever they congregate, like members of a country sewing society.

And yet these wary creatures have been successfully domesticated and crossed with the common barnyard goose. Many stories are in circulation of wild geese that have been wounded, and placed among the farmer's fowls, where they have been made well and apparently content until a flock of migrants, passing above, called them to a wild life again; but the very birds that could be easily identified by the scars of old wounds, revisited the barnyard whenever their travels to and from the south permitted. All geese become strongly attached to certain localities. Ordinarily, a goose that has been wounded in the wing runs, if on land, but so awkwardly it may be quickly overtaken. If wounded when above or on the water, it will dive, and remain under the surface with only its nostrils exposed until all danger is over. Unless seriously hurt, it generally eludes capture. The thick coat of feathers, that have an even greater commercial value than its flesh, is the goose's suit of armor, impenetrable except at close range.

When surprised, a flock rises suddenly in great confusion; the large birds get in one another's way and offer the easiest shots the tyro ever gets; the *honk, honk, k'wonk* from many outstretched throats clamoring at once mingles with the roar of wings, as with slow, heavy, labored flight the geese rise against the wind—the point from which they must be approached if one

Geese

is to get a good view of them. But order somehow comes speedily out of chaos once the birds are well launched in air. Double ranks are formed, with the leader at the point where the two lines converge, and the wedge moves on, far away if they have been terrorized by firing, but only a few hundred yards if they find there is no real ground for fear.

Flocks of wild geese go and come in the United States from September, when the young birds are able to join in the long flights, until early spring, when the great majority go north to nest. In some secluded marsh, by the shores of streams, or on the open prairie, far from the habitations of hungry men, the goose lays four or five pale buff eggs in a mass of sticks lined with grass and feathers, and sits very closely, while the gander keeps guard near by. An empty osprey's nest in a tree top, or a cavity in some old stump, frequently contains these eggs; but the goslings never return to the cradle once they have been led to water, for they are good walkers and swimmers from the start. After a thorough moult, which often makes the old birds as incapable of flying as the goslings, the detached families gather into flocks in September, when a few cold snaps in the Hudson Bay region suggest the necessity for migrating to warmer climes. On their arrival here they are very thin, worn out by the long journey; but the Christmas goose, as every housekeeper knows, is perhaps the fattest bird brought to her kitchen.

Brant

(Branta bernicla)

Called also: BRENT; BRANT GOOSE; AND BARNACLE GOOSE

Length—26 inches.

Male and Female—Head, neck, throat, and upper breast and shoulders blackish, with a small patch of white streaks on either side of neck, sometimes also on chin and lower eyelid; back brownish gray, the feathers margined with ashy; lower breast ashy gray, ending abruptly at the line of black of the upper breast; sides dark, but fading into white underneath; much white around tail; bill and feet black. Female smaller than gander. Immature birds have no white patch on neck, and plumage above and below is barred or waved with reddish brown.

Range—Arctic sea, nesting within the Arctic Circle, to the Carolinas in winter. Most common on Atlantic coast; rare in the interior.

Season—Winter resident, or spring and autumn migrant in the United States.

Flocks of brants continue to fly southward down the Atlantic coast from October until December, some alighting on muddy flats around the estuaries of rivers and creeks, on sand bars and in shallow inlets, to feed on eel-grass and other marine plants; but the majority passing rapidly by the shores of Canada and our northern states. High flyers, sea lovers, they keep well out from land during the migrations rather than follow the coast line, if any distance may be saved by a bee line from point to point. It is only in hazy weather that they fly low. A reconnoitre by the veterans must first be made after the confused mass of hoarse gabblers rises from the feeding grounds; but after this spiral soaring has ended and the birds are once fairly started on their journey, neither pause nor uncertainty may be detected in their steady flight. They fly in more compact bodies than the long-drawn-out wedges of Canada geese; no leader appears to direct their course, yet the mass moves as one bird, slowly and sedately. Some one has compared the trumpet-like sounds made by a flock of brants with the noise of a pack of fox-hounds in full cry. Occasionally these geese are found in the interior, for all their strong maritime preferences; but usually it is the black brant that is mistaken for them there and on the Pacific slope.

On Long Island and southward these dusky waders walk about at low tide, tearing up eel-grass by the roots when they enter the marshes to feed in gabbling, honking companies. Watched from a distance—for a close approach, no matter how stealthy, frightens these wary birds to wing—they appear rather sluggish and move heavily over the mud flats, nipping every plant that grows in their path. Youthful gunners constantly mistake them for some of the larger sea-ducks and wonder that they do not dive for food. Brants never dive unless wounded. While the tide is out they feed constantly, stopping only to gabble and gossip, and quarrel from excessive greediness, with the result of being too heavy and lazy with much gorging to fly out to sea when the tide comes in and lifts them off their feet. After sundown they go streaming in long lines out to deep, open

Geese

water to pass the night afloat. Certain localities become favorite stopping places for these birds of passage, and they return to them year after year, unless harassed by the gunners beyond endurance; but such resorts become rarer every season. In early winter the young of the year are as delicious a game bird as finds its way to the gunner's pouch; but old birds taken in the spring migration defy the inroads of any tooth not canine.

Because it nests so very far to the north, the life history of this goose is still incomplete. According to Saunders, the nest is composed of grasses, moss, etc., lined with down and made on the ground. Four smooth and creamy white eggs fill it.

. . . .

The Black Brant *(Branta nigricans)*, a name sometimes applied to the white-fronted goose to distinguish it from the white brant or snow goose, is the western representative of the preceding species and of only casual occurrence on the Atlantic coast. It may be readily distinguished from its ally by its darker under parts and the white markings on the front as well as the sides of its neck. Their habits are almost identical. Both these "barnacle geese" take their name, not from their fondness for the little crustacean, for they are almost vegetarians, but from the absurd fable that they grew out of barnacles attached to wood in the sea. Some etymologists claim that the word brant is derived from the Italian word *branta*, coming from *branca*, a branch; but these geese have nothing to do with branches, unlike the Canada geese that sometimes nest in trees; and we may more confidently accept Dr. Coues's statement that brant means simply burnt, the dark color of the goose suggesting its having been charred.

SWANS

(Subfamily Cygninæ)

Whistling Swan
(Olor columbianus)

Called also: AMERICAN SWAN

Length—55 inches, or a little under 5 feet.

Male and Female—Entire plumage white; usually a yellow spot between the eyes and nostrils, but sometimes wanting; bill, legs, and feet black. Immature birds have some brownish and grayish washings on parts of their plumage.

Range—North America, nesting about the Arctic Ocean, and migrating in winter to our southern states and the Gulf of Mexico. Rare on the Atlantic coast north of Maryland; more abundant on the Pacific.

Season—Winter visitor and spring and autumn migrant, October to April.

It is impossible for one who has seen only the common mute swans floating about in the artificial lakes of our city parks, while happy children toss them bits of cake and crackers, to imagine the grandeur of a flock of the great whistlers in their wild state. Not far from Chicago such a flock was recently seen in its autumn migration, and as the huge birds rose from the lake into the air, it seemed as if an aërial regatta were being sailed overhead; the swans, each with a wing-spread of six or seven feet, moving like yachts under full sail in a mirage where water blended with sky and tricked one's vision. The sight is among the most impressive in all nature. It is wonderful!

On the Pacific coast, in the interior, down the Mississippi to the gulf states, and up the Atlantic coast from Florida to the Chesapeake, the whistling swans wander between October and April, flying at the rate of one hundred miles an hour, it is

Swans

estimated. Like many of their smaller relatives, they fly in wedge shaped flocks, with an experienced, clarion voiced veteran in the lead. Dr. Sharpless, who was the first to point out this species as distinct from the whooping or whistling swan of Europe, with which our early ornithologists confused it, says: "Their notes are extremely varied, some closely resembling the deepest bass of the common tin horn, while others run through every modulation of false note of the French horn or clarionet." The age of the bird is supposed to account for the difference in the voice. No one can mistake the notes for the product of any musical instrument, however. One unkind man in the south, who was wakened in the depth of night by the noisy trumpetings of a flock feeding in a lagoon near his home, was heard to remark that if the swan did not really sing just before its death, it really ought to die just after making that noise! The poets, from Homer to Tennyson, and not the scientists, are responsible for the story of the swan's chanting its own dirge. These swans are particularly noisy when dressing their feathers, when feeding, and when flying, especially just after mounting from the water into the air, when they make loud demands each for its proper place in the V-shaped column. The Indians say that the swans follow in the wake of a flock of geese. Perhaps the Hudson Bay Fur Company, which has bought thousands of pounds of swan's down from the Indians, best knows why there are so few flocks of swans left to follow the geese to-day.

Around the shores of lakes and islands in the Hudson Bay region, these swans return to nest in May; and gathering a mass of sticks and aquatic plants, pile them to a height of two feet or more, this down-lined nest being sometimes six feet across. In the labor of making it the male helps, for he is a far better mate and father than either a drake or a gander. From two to six rough, grayish eggs, over four inches long and nearly three inches wide, are laid in June, and not until after five weeks of close confinement on the nest can the proud mother lead her brood to water. At first the fledgelings are covered with a grayish brown down, which gradually changes into the white plumage that it takes twelve months to perfect. Young cygnets are counted a great delicacy by the epicures of Europe.

Had the prehistoric swans been content to nibble herbage on the banks of streams, instead of immersing their necks to probe the bottoms for mollusks, worms, and roots, doubtless their necks would have reached no abnormal length. One rarely sees a swan tipping after the manner of the river ducks, and never diving. To escape pursuit the swan, which is really very shy, will quickly distance a strong rower by swimming, yet with an ease and majesty of movement that suggests neither fright nor haste.

. . . .

The Trumpeter Swan *(Olor buccinator)*, an even larger species than the preceding, with no yellow on the fore part of its head, though elsewhere identical in plumage with the whistler, has a more western range, being rarely found east of the Mississippi. In habits the two great birds appear to be much the same, but the voice of the well-named trumpeter resounds with a power equalled only by the French horns blown by red-faced Germans at a Wagner opera.

PART II
WADING BIRDS

HERONS AND THEIR ALLIES

Ibises
Storks
Bitterns
Herons
Egrets

HERONS AND THEIR ALLIES
(Order *Herodiones*)

Spoonbills, herons, storks, bitterns, ibises, flamingoes, egrets, or white herons, and their kindred compose an order remarkable for the large average size of its members, all of whom have either long legs or necks, or both. Most of these birds belong to the tropics; and while many of them formerly reached our southern states in great numbers, the greed of the plume hunter, incited by the thoughtless vanity of women, has nearly exterminated a number of the most beautiful species. The majority of these birds are either local or have now become too rare to be included in this book.

Ibises
(Family *Ibididæ*)

Slender, picturesque birds, long of neck, bill, legs, and wings, and very short tailed. A bare space around eye; claws almost like human nails. Silent birds, always living in flocks, chiefly on shores of smaller bodies of water or on bars and lower beaches on which the outgoing tide leaves a harvest of small crustaceans, which with frogs, lizards, small fish, etc., form their food. Sexes alike ; young different.

White Ibis, or Spanish Curlew.

Storks and Wood Ibises
(Family *Ciconiidæ*)

Unhappily these storks still retain the name "ibis," which no amount of scientific protest seems possible to shake off. General form as in preceding group; but bill, which is as broad as the face at base, has tip curved downward. Four long toes, the hind one about on the level with the front ones, enabling the birds to rake the muddy bottoms of shallow lagoons with their feet. Claws less nail-like than in true ibises. Strong, graceful fliers.

Wood Ibis

Herons and Bitterns
(Family Ardeidæ)

Birds of this family, that contains about seventy-five species, mostly confined to the tropics, have certain peculiar feathers or "powder-down tracts" which, when worn in pairs of two or three, are a fair but superficial mark of the clan. The herons wear three pairs; one on the back, over the hips; one underneath the hips, on the abdomen; and another on the breast. Bitterns lack the pair underneath. Their purpose is not yet known, but some scientists contend that these tracts are phosphorescent, and that fish are lured by them at night. The plumage is generally loose, adorned with lengthened feathers, some species having beautiful crests and plumes on the back, that are worn in the nesting season. The legs are long and unfeathered, for wading; the four toes, all on the same level, are long and slender, for perching. The bill, which is always longer than the elongated, narrow head, appears to run directly into the eyes. Usually herons nest and roost in flocks, in favorable localities, numbering thousands; but when feeding on the shores of lagoons, rivers, and lakes, solitary birds are seen. Other species, on the contrary, live singly or in pairs all the time.

American Bittern, or Marsh Hen.
Least Bittern.
Great Blue Heron, or Blue Crane.
Little Blue Heron, or Blue Egret.
Snowy Heron, or White Egret.
Green Heron, or Poke.
Black-crowned Night Heron, or Quawk.

IBISES
(Family Ibididæ)

White Ibis
(Guara alba)

Called also: SPANISH CURLEW

Length—25 inches.

Male and Female—Plumage white, except the tips of four outer wing feathers, which are black. Bare space on head; most of bill and the long legs orange red. Long decurved bill tipped with dusky. Immature birds dull brown, except lower back and under parts, which are white.

Range—Warmer parts of United States, nesting as far north as Indiana, Illinois, and South Carolina; straying northward annually to Long Island, and casually to Connecticut and South Dakota; winters in West Indies, Central, and northern South America.

Season—Summer resident or visitor.

Flocks of these stately, picturesque birds, flying in close squadrons, their plumage glistening in the glare of a tropical sun, their legs trailing after them, are not so familiar a sight even in the Gulf states as once they were. Their destruction can be set down to nothing but wanton cruelty, for their flesh is totally unfit for food, and their usefulness is *nil* if it does not consist in enlivening waste places with their beauty.

Morning and evening the close ranks fly to and from the feeding grounds on the shores of lagoons and lakes, or to their favorite roosts, where their ancestors likely as not slept before them. Standing on one leg, with head and bill drawn in to rest between the shoulders and on the breast, the body in a perpendicular position, an ibis can remain motionless for hours, a picture of tropical indolence. The bill, which so closely resembles the curlew's that this ibis is frequently called Spanish cur-

Ibises

lew, enables the bird to drag out the crayfish from its shell and pinch the last piece of flesh from soft-shelled crustaceans. Small fish, frogs, lizards, and other aquatic animal food never seem to fatten this slender bird, that is a ravenous feeder none the less.

Colonies of ibises build nests in ancestral nurseries, which may be in reedy marshes, or in low trees and bushes not far from good feeding grounds. Three to five pale greenish eggs marked with chocolate are found in the coarse, bulky nest of reeds and weed stalks.

STORKS AND WOOD IBISES
(Family *Ciconiidæ*)

Wood Ibis
(*Tantalus loculator*)

Called also: WOOD STORK; COLORADO TURKEY; WATER TURKEY

Length—40 inches.

Male and Female—Head and neck bare, and bluish or yellowish; plumage white, except the primaries and secondaries of wings and the tail, which are greenish black. Legs blue, blackish toward the toes; long, thick, decurved bill, dingy yellow. Immature birds have head covered with down; plumage dark gray, with blackish wings and tail, but soon whitening.

Range—"Southern United States, from the Ohio Valley, Colorado, Utah, southeastern California, etc., south to Argentine Republic; casually northward to Pennsylvania and New York."—A. O. U.

Season—Resident, or summer visitor.

Like the turkey buzzards, this wood stork has the fascinating grace of flight that one never tires of watching, as the birds, first mounting upward with strong wing beats, go sailing away overhead in great spirals, floating on motionless, wide wings, wheeling, gyrating, rising, falling, skimming in and out of the pathless maze that a flock follows as if its members were playing a sedate game of cross tag. With necks distended and legs trailing on a horizontal with their bodies, their length is extreme. As these birds are gluttonous feeders, it has been suggested that their flights, like the buzzard's, are taken for exercise to quicken their digestion.

There is a tradition to the effect that the wood ibis is a solitary misanthrope, but Audubon mentions thousands in a flock; and while the day of such sights has passed forever in this land of bird butchers, one rarely sees a lone fisherman in the south

Storks and Wood Ibises

to-day, and where one meets the bird at all, it is likely to be in the company of at least a score of its kind, with possibly a few buzzards sailing in their midst. "The great abundance of the wood ibis on the Colorado, especially the lower portions of the river," says Dr. Coues, "has not been generally recognized until of late years, . . . but the swampy tracts and bayous of Louisiana, Mississippi, Alabama, and Florida are . . . their favorite homes."

Speaking of a hunting trip on the Myakka River in west Florida, in 1879, Mr. G. O. Shields writes: "As we walked quietly around a bend in the river, just out of sight of our camp, and came to an open glade or meadow of perhaps an acre, a sight met our eyes that might have inspired the soul of a poet or have awakened in the mind of the prosiest human being visions of Paradise. There sat great flocks of richly colored birds, the backs of which were nearly white, the wings and breast a rich and varied pink, changing in some of the males to almost scarlet. These were the roseate spoonbills [now nearly extinct]. In another part of the glade was a large flock of the stately wood ibis, with body of pure white, and wings a glossy radiant purple and black. In still another part, a flock of snowy white egrets, and here and there a blue or gray heron, or other tropical bird. Alarmed at our approach they all arose, but, as if aware their matchless beauty was a safeguard against the destroying hand of man, they soared around over our heads for several minutes before flying away. As they thus hovered over us we stood and contemplated the scene in silent awe and admiration. Our guns were at a parade rest. We had no desire to stain a single one of the exquisite plumes with blood."

Indolent as creatures of the tropics are wont to be, the wood stork goes to no further effort to secure a dinner than dancing about in the shallow edges of the lagoon, to stir up the mud, which brings the fish to the top. A sharp stroke from its heavy bill leaves the fish floating about dead to serve as bait. With head drawn in between its shoulders, a pensive, sedate figure, the stork now calmly waits for other fish, frogs, lizards, or other reptiles to approach the bait, when, quick as thought, it strikes right and left, helping itself to the choicest food, and leaving the rest for the buzzards and alligators. A sun bath after such a gorge completes its happiness.

HERONS AND BITTERNS

(Family Ardeidæ)

American Bittern

(Botaurus lentiginosus)

Called also: MARSH HEN; INDIAN HEN; STAKE DRIVER; POKE; FRECKLED HERON; BOG BULL; NIGHT HEN; BOOMING BITTERN; LOOK-UP

Length—Varies from 24 to 34 inches.

Male and Female—Subcrested; upper parts freckled with shades of brown, blackish, buff, and whitish; top of head and back of neck slate color, with a yellow-brown wash; a black streak on sides of neck; chin and throat white, with a few brown streaks; under parts pale buff, striped with brown; head flat. Bill yellow, and rather stout, and sharply pointed; tail small and rounded; legs long and olive colored.

Range—Temperate North America; nests usually north of Virginia, and winters from that state southward to the West Indies.

Season—Summer resident, or visitor from May to October; permanent in the south.

The booming bittern, whose "barbaric yawp" echoes from lonely marshes, grassy meadows, and swamps through the summer, enjoys greater popularity in name than in deed; for he is a hermit, a shy, solitary wanderer, that even Thoreau, no less secluded than he, knew by his voice chiefly. "Many have heard the stake driver," says Hamilton Gibson, "but who shall locate the stake?" The same bird whose voice sounds like a stake being driven into a bog, or, again, "like the working of an old-fashioned wooden pump," or like the hoarse crowing of a raven when it flies at night, has for its love song the most dismal, hollow bellow, that comes booming from the marshes at evening, a mile away, with a gruesome solemnity. One of these calls has been written *pump-er-lunk, pump-er-lunk, pump-er-lunk;* but a better rendering, perhaps, is Dr. Abbott's *puck-la-grook*, which has been verified again and again.

After the sedges in the marshes have grown tall, it is next to impossible to find the bird; but on its arrival in spring, when it pumps most vociferously in the fens, the paddler up some lonely creek follows the sound until he sees this freckled fellow standing perfectly still in the low grass, its head held erect and pointed upward. Not a muscle moves while the bird remains in ignorance of the watcher. An hour passes, and it might be a dead stump standing there in the twilight. It looks particularly like a stump if it has assumed another favorite position, of drawing in its head until it touches its back. Suddenly a succession of snappings and gulpings, to fill its lungs with air, convulses the creature, and then three booming bellowings come forth with gestures that suggest horrible nausea. One who did not see the bird in the act of making these noises would imagine from their quality that they came from below the water, and there are many stories in circulation among people who do not go to the pains to verify them, that water is actually swallowed and ejected by bitterns to assist their voices; but it is not.

Come upon the hermit suddenly, and it seems paralyzed by fright. When danger actually threatens, up go the long head feathers, leaving the neck bare and making the bird look formidable indeed. The plumage is ruffled, the wings are extended, and if the adversary comes too near, a violent slap from the strong wing and a thrust from the very sharp beak makes him wish his zeal for bird lore had been tempered with discretion. A little water spaniel was actually stabbed to death as a result of its master's inquisitiveness.

During the day, the bittern, being extremely timid, keeps well hidden in the marshes; but it is not a nocturnal bird, by any means, however well it likes to migrate by night. To some it may appear sluggish and indolent as it stands motionless for hours, but it is simply intelligently waiting for frogs, lizards, snakes, large winged insects, meadow mice, etc., to come within striking distance, when, quick as thought, the prey is transfixed. A slow, meditative step also gives an impression of indolence, but the bittern is often only treading mollusks out of the mud with its toes.

LEAST BITTERN.
½ Life-size.

In the air the bittern still moves slowly, and with a tropical languor flaps its large, broad wings, and trails its legs behind, to act as a rudder as it flies close above the tops of the sedges. When a longer journey than from one part of the marsh to another must be made, the solitary traveller mounts high by describing circles; and, secure under the cover of darkness, makes bold and long excursions. It is only in the nesting season that we find these birds in couples. Then neither one is ever far away from the rude grassy nest that holds from three to five pale olive buff eggs hidden among the sedges, on the ground, in a marsh. There are those who assert that young bitterns are good food.

Least Bittern
(Ardetta exilis)

Called also: TORTOISE-SHELL BIRD; LITTLE BITTERN; FLY-UP-THE-CREEK

Length—13 inches.

Male—Subcrested; top of head, back, and tail black, with green reflections; back of neck and sides of head brownish red, also wings, coverts, and edges of some quills; throat whitish, shading into buff on under parts; the deepest shade, almost a yellow-brown, on sides; much buff on wings. Bill, eyes, and feet yellow; legs long and greenish.

Female—Similar to male, but chestnut above, and the darker under parts are lightly streaked with dark brown.

Range—Throughout temperate North America, nesting from Maine and the British Provinces southward; winters from Gulf states to West Indies and Brazil; less common west of the Rocky Mountains, but found on the Pacific coast to northern California.

Season—Summer resident.

The smallest member of a family of waders noted for their large size, the least bittern brings down their average considerably; for it is only about a foot long, a quarter the length of the next species. Fresh-water marshes, inaccessible swamps, boggy lands, and sedgy ponds are where these secretive little birds hide, with rails and marsh wrens, gallinules, bobolinks, red-winged blackbirds, and swamp song sparrows for neighbors among the rushes. Living where no rubber boot may follow them through the muck, they usually remain unknown to many human neighbors, unless some sluggish stream running through their territory will float a skiff and a bird student within field-glass range. These bitterns are by no means the solitary hermits the larger species are. Colonies of a dozen or more couples are found nesting within the same acre.

However retiring in habits by preference, the least bitterns show no especial shyness when approached. Mr. Chamberlain tells of a small colony that spend the summer within a stone's throw of a street-car track and a playground in the busiest part of Brookline, near Boston—probably the home their ancestors were reared in; for all the birds of this family show marked respect and attachment for an old homestead. In Westchester County, New York, there is a certain sluggish river whose reedy shores contain twenty nests or more within sight of a well-worn foot bridge. Here, looking down into the sedges, the birds are seen running about through the jungle, with their necks outstretched and their heads lowered, as they hunt for food—small minnows, or young frogs and tadpoles, lizards, and bugs winged and crawling. Disturb the birds, and they take wing at once, with a harsh, croaking note. *qua*, and flapping their wings slowly and heavily, retreat no farther than to a denser part of the marsh, into which they drop, and are lost in the rushes.

Dr. Abbott writes of a bittern's nest that he found near Poaetquissings Creek—that mine of nature's treasures he has opened for the delight of easy-chair naturalists. "Such finds make red-letter days," he says. "The nest itself was a loosely woven mat of twigs and grass, yet strong enough to be lifted from the tuft of bulrush upon which it rested. There were a single dirty blue white egg and four fuzzy baby bitterns not a week old. They were clad in pale buff down, scantily dusted over them, and an abundance of straight white hairs as long as their bodies. These young birds were far less awkward, even now, than herons of the same or even greater age. As I took one up, it thrust its opened beak at me, but, becoming quickly reconciled, seemed to take pleasure in the warmth of my hand. At times it uttered a peculiarly clear, fifelike cry . . . free from every trace of harshness."

Near sunset and in the twilight of night and morning is

when these bitterns, like all their kin, step boldly out of their retreats and indulge in longer flights from home. Many men of science have thought the powder-down tracts on their bodies glow with phosphorescent light in the dark and attract fish to the water's edge, where the bird stands motionless, ready to transfix a victim with its beak. But as yet this is only an interesting theory that has still to be proved.

Great Blue Heron
(Ardea herodias)

Called also: BLUE CRANE, (erroneously) SANDHILL CRANE

Length—42 to 50 inches. Stands about 4 feet high.

Male and Female—Crown and throat white, with a long black crest beginning at base of bill, running through eye, and hanging over the neck, the two longest feathers of which are lacking in autumn. Very long neck, light brownish gray, the whitish feathers on lower neck much lengthened and hanging over the dusky and chestnut breast. Upper parts ashy blue; darker on wings, which are ornamented with long plumes, similar to those on breast, in nesting plumage only. Bend of wing and thighs rusty red. Under parts dusky, tipped with white and rufous. Long legs and feet, black. Bill, longer than head, stout, sharp, and yellow.

Range—North America at large, from Labrador, Hudson Bay, and Alaska; nesting locally through range, and wintering in our southern states, the West Indies, and Central and South America.

Season—Summer resident at the north, April to October, often to December; elsewhere resident all the year.

The Japanese artists, "on many a screen and jar, on many a plaque and fan," have taught some of us the æsthetic value of the heron and its allies—birds whose outstretched necks, long, dangling legs, slender bodies, and broad expanse of wing give a picturesque animation to our own marshes. But American artists seek them out more rarely than shooters, and a useless mass of flesh and feathers lies decomposing in many a morass where the law does not penetrate and the rifle ball does. Longfellow, in "The Herons of Elmwood," paints a word picture of

Herons and Bitterns

this stately bird, full of appreciation of its beauty and the mystery of the marsh. Surely no one enjoyed

> "The cry of the herons winging their way
> O'er the poet's house in the elmwood thickets"

more than Lowell himself.

> "Sing him the song of the green morass,
> And the tides that water the reeds and rushes;
> . . . .
> Sing of the air and the wild delight
> Of wings that uplift and winds that uphold you,
> The joy of freedom, the rapture of flight
> Through the drift of the floating mists that infold you."

Hern, an obsolete form of heron, was perhaps last used by Tennyson when he wrote of "The Brook" that comes "from the haunts of coot and hern." The old adage, "not to know a hawk from a handsaw," lacks its meaning if we do not recall how heronsewe, a heron (not heronshaw, as is often written), was corrupted in England long ago, when hawking was a favorite sport there, into hernser, in turn corrupted into handsaw. Tradition says that the soul of Herodias became incarnate in the heron, the favorite bird of Herod, but in that case the common heron of Europe (*Ardea cinerea* of Linnæus) should bear her hated name, and not this distinctly American species.

Patience, an easy virtue of the tropics, from whence the great blue heron comes, characterizes its habits when we observe them at the north. Standing motionless in shallow water, the Sphinx-like bird waits silently, solemnly, hour after hour, for fish, frogs, small reptiles, and large insects to come within range; then, striking suddenly with its strong, sharp bill, it snaps up its victim or impales it, gives it a knock or two to kill it if the thrust has not been sufficient, tosses it in the air if the prey is a fish, and, in order to avoid the scratching fins, swallows it head downward. Hunters pretend to excuse their wanton slaughter by saying herons eat too many fish; but possibly these were created as much for the herons' good as our own, and no thanks are offered for the reptiles and mice they destroy.

Wild, shy, solitary, and suspicious birds, it is next to impossible to approach them, even after one has penetrated to the forbidding retreats where they hide. Near sunset is the hour

GREAT BLUE HERON.
⅕ Life-size.

they prefer to feed. In Florida one meets herons constantly, fishing boldly on the beach, wading in the lagoons, perching on stumps, and walking with stately tread and slow through the sedges by the river side, their long necks towering above the tallest grasses. The cypress swamps all through the south contain herons of every kind; but at the north the sight of this lone fisherman is rare enough to be memorable. Nine times out of ten he will be standing with his head drawn in to rest between his shoulders, and motionless as a statue. As he generally chooses to fish under the shadow of a tree by the water, or among the rushes that grow out into the sluggish stream, his quiet plumage and stillness protect him from all but the sharpest eyes. Disturb him, and with a harsh rasping *squawk* he spreads his long wings, flaps them softly and solemnly, and slowly flies deeper into the marsh. At close range he looks a comical mass of angles; but as he soars away and circles majestically above, his great shadow moving over the marsh like a cloud, no bird but the eagle is so impressively grand, and even it is not so picturesque.

Herons are by no means hermits always. Colonies of ten or fifteen pairs return year after year at the nesting season to ancestral rookeries, each couple simply relining with fresh twigs the platform of sticks in a tree top that has served a previous brood or generation as a nest. The three or four dull bluish green eggs that are a little larger than a hen's very rarely tumble out of the rickety lattice, however. Both the crudeness of the nest and the elliptical form of the egg indicate, among other signs, that the heron is one of the low forms of bird life, not far removed, as scientists reckon space, from the reptiles. Sometimes nests are found directly on the ground or on the tops of rocks; but even then the fledgelings, that sit on their haunches in a state of helplessness, make no attempt to run about for two or three weeks.

. . . . . .

The Little Blue Heron, or Blue Egret *(Ardea cærulea)*, less than half the size of its great cousin, casually wanders northward and beyond the Canadian border when its nesting duties are over in southern rookeries. Its home is also a platform of sticks, but it is placed, with a dozen or more like it, in bushes over the watery hunting ground, and not in the tops of tall cypresses or other trees. Such colonies are still found as far north as Pennsylvania and southern Illinois. A rich maroon brown head and neck set off its bluish slate plumage, which is adorned with lengthened pointed feathers on the breast and shoulders. Immature birds are more confusing. At first they are white, or white washed with slaty gray, the tips of the primaries always remaining bluish slate, however, which enables one to tell them, with the help of their greenish yellow legs, from the snowy herons or egrets so often confused with them. Happily, the little blue herons wear no aigrettes, or they would share the tragic fate of the beauty of their family.

* * * * *

> "What does it cost, this garniture of death?
> It costs the life which God alone can give;
> It costs dull silence where was music's breath;
> It costs dead joy that foolish pride may live;
> Ah, life and joy and song, depend upon it,
> Are costly trimmings for a woman's bonnet!"

Only a generation ago the Snowy Heron (*Ardea candidissima*) was so abundant the southern marshes fairly glistened with flocks, as if piled with snow; but all the trace of this exquisite bird now left is in the aigrettes that, once worn as its wedding dress, to-day wave above the unthinking brows of foolish women. In some states there is a penalty attached to the shooting of this heron; but the plume hunters evade the law by cutting the flesh containing the aigrettes from the back of the living bird, that is left to die in agony. Countless thousands of the particularly helpless fledgelings, suddenly orphaned, have slowly starved to death, and so rapidly hastened the day when the extinction of the species must end the sinful folly.

Little Green Heron

(Ardea virescens)

Called also: POKE; CHUCKLE HEAD; CHALKLINE; FLY-UP-THE-CREEK

Length—16 to 18 inches; smallest of the herons.
Male and Female—Lengthened crest and crown of head dark green; rest of receding head and neck chestnut red, shading

into yellow; brownish ash under parts; throat white, with line of dark spots widening on breast; back, with pointed lengthened feathers between shoulders, is green, or washed with grayish; wings and tail dark green, the coverts of the former outlined with white. Bill long and greenish black. Rather short legs, greenish yellow. Immature birds lack the lengthened feathers on back, are less brilliant, their crests are smaller, and they have black streaks on their under parts.

Range—Tropical and temperate America; nests throughout the United States and far into the British possessions; winters from Gulf states southward.

Season—Summer resident, April to October.

This smallest, most abundant, and most northern heron comes up from the south in lustrous green plumage that gradually loses its iridescence as nesting duties tell upon the physique; but as it is a solitary, shy bird, very few get a close look at its feathers at any time. Delighting in quagmires, where no rubber boot stays on the foot of the pursuer, the little green heron goes deeper and deeper into the swamp, and keeps well concealed among the rushes by day, coming out to the shores of wooded streams and sedgy ponds toward dusk, when often as not the motionless little figure is mistaken for a snag and passed by.

Not a muscle does the bird move while patiently waiting for fish, frogs, and newts to come within striking distance of its sharp bill. With head drawn down between its shoulders, it will stand motionless for more hours than the most zealous bird student cares to spend watching it. Where food is exceedingly abundant, one may sometimes be seen wading around the edge of the pond with slow, well calculated steps, snapping up the little water animals that also become more active as evening approaches.

Startle the lone fisherman, and with a hollow, guttural *squawk* it springs into the air, but does not flap its wings long before dropping on some old stump or distended branch to learn whether further flight is necessary. There is a certain laziness or languor about all the herons that they have brought from the tropics with them. When perched on a stump, its receding head thrust forward like a stupid, its apology for a tail twitching nervously, one sees the fitness of many of this heron's popular names. But why is this inoffensive wader held in such general contempt?

It has been stated by some scientists that, unlike many of its kin, the green heron is always a hermit, rarely seen in couples, and never found in colonies, even at the nesting season; but surely there are enough exceptions to prove the rule. From all points of its large nesting range come accounts of heronries where not only green herons have built their rickety platforms of sticks in the low branches of trees or bushes in communities, but have associated there with different relatives, particularly with the night heron. They begin to build nests, or reline what the winter storms have left of their old ones, about the middle of April. These birds become attached to their nesting sites that they return to generation after generation, and a roost often becomes equally dear. There are certain favorite trees in localities where the green heron is abundant that one rarely misses finding a bird perched upon.

Why it is that the eggs—pale dull blue, from three to six—and the helpless fledgeling do not fall out or through their ramshackle nursery is a mystery. Indolence characterizes these birds from infancy; for they remain sitting on their haunches in a state of inertia, roused only by visits of their enslaved parents bringing them food, until they are perfectly able to fly, some weeks after hatching.

Black-crowned Night Heron

(Nycticorax nycticorax nævius)

Called also: QUAWK; QUA BIRD

Length—23 to 26 inches. Stands fully 2 feet high.

Male and Female—Three long white feathers, often twisted into apparently but one, at the back of head, worn only at the nesting season. Crown and back greenish or dull black; wings, tail, and sides of neck pearl gray with a lilac tint; forehead, throat, and underneath white. Legs and feet yellow; eyes red; bill stout and black. Immature birds very different: grayish brown, streaked or spotted with buff or white on upper parts; under parts white streaked with blackish; some reddish brown feathers in wings.

Range—United States and British provinces, nesting from Manitoba and New Brunswick southward to South America. Winters in Gulf states and beyond.

Season—Summer resident, or spring and autumn migrant north of the southern states. Resident all the year at the south.

To say that this is the most sociable member of a family that contains many misanthropic hermits, gives little idea of the night heron's fondness for society. Colonies of hundreds of pairs are still to be found, thanks to the bird's secluded and nocturnal habits. Some heronries contain these birds living among the blue, the great blue, or the green species, but in no very advanced state of socialism, however, for the gossiping and noisy *quawking* over petty quarrels that constantly arise make the place a pandemonium. Wilson, who usually pays only the kindest, most appreciative compliments to birds, likens the noise made by these to that of two or three hundred Indians choking each other!

Not because the flesh of this bird is good for food, or its plumage is desired for hats, but because it is a nuisance in the neighborhood where civilization creeps upon the ancient eyries, is the night heron hunted. Flocks become so attached to the home of their ancestors, that only the harshest persecution drives them away, and then often no further than a few hundred rods. A sickening stench pervades the air blowing off a heronry; decomposed portions of fish, frogs, mice, and other animal food lie about on the ground, that is white with the birds' excrements. At Roslyn, Long Island, almost within sight of New York, a large colony of night herons that were driven from a populated portion of the town, where they had nested and roosted for many years, finally settled in a well wooded swamp not far off only after disgraceful persecution. One man boasts of having shot three hundred. Nevertheless there must be a thousand birds there still. For their protection, it should be added that there are few less inviting places to visit on a summer's day than this heronry. Certainly there is as much sport in shooting at the broad side of a barn as in hitting one of these large birds that, dazed by the sunlight, sits motionless on a distended branch, where any tyro could hit it blindfolded.

The night herons arrive from the south about the middle of April, and at once repair what is left of the rickety platforms of sticks used a previous season, or build new ones. The wonder is they can weave any sort of a lattice out of such stiff, unyielding material. These nests are generally in the tops of tall trees, especially the cypresses, swamp oaks, and maples and evergreens near or growing out of a swamp; but there are also records of nests in bushes, or even on the ground. Often fluffy, helpless fledgelings are found climbing about the nest while there are still some dull, pale blue eggs unhatched in June, which suggests the possibility of the extension of socialism into the nurseries; but who knows whether the rightful parents rear only their own young?

Toward sunset all the eyries in the swamps are emptied, and although, while the broods are young and incapable of making any effort whatever, the old birds must go a-fishing by day as well as at dusk, it is at twilight and later in the night that these herons choose to disperse among the ditches, shores of ponds and streams, the bogs and marshy meadows, to gorge upon the teeming animal life there. Next to this bird's fondness for an old, colonial homestead, its insatiable appetite is perhaps its most prominent characteristic. Evidently the digestion of a young heron keeps in a state of perpetual motion. The old birds, slender as they always are, grow perceptibly thinner while raising their two broods a year. A choking noise, like the painful effort to bring up a fish that has taken a wrong course down the bird's long throat, but which is only an attempt to sing or converse, that old and young alike are constantly making, keeps a heronry well advertised, much to the profit of the hawks.

Standing motionless, with head drawn in between its shoulders, as it waits at the margin of a pond at evening for the food to come within striking range, the heron can scarcely be distinguished from a crooked stick. However deficient its sight may be, especially by day, an extraordinary keenness of ear detects the first creak of an intruder's foot, and with a *quawk, quawk,* the bird rises and is off, trailing its legs behind, after the manner of storks that Japanese artists have made so familiar.

Have birds a color sense? A night heron that was seen perching among the gray branches of a native beech tree must have known how perfectly its coat blended with its surroundings, where it was all but invisible to the passers by.

BLACK-CROWNED NIGHT HERON

MARSH BIRDS

Cranes
Rails
Gallinules
Coots

MARSH BIRDS

(Order *Paludicolæ*)

Birds of the plains and marshes, the two families comprising this order have certain resemblances of structure that unite them into a distinct order, however the large cranes, with their long necks and legs, seem to approach more nearly the herons and their allies than they do the small rails, or marsh hens, and their congeners. Cranes, rails, gallinules, and coots, unlike the altricial heron tribe, are precocial; that is, they run at once from the nest, well clothed with down when hatched. These birds have four toes, the three front ones long, to enable the birds to run lightly over the oozy ground; the hallux, or great toe, may be elevated at the back, or, as in the case of gallinules and coots, on the level with the front ones, which in several species are lobate, but not flattened also, or palmate, as the grebes' toes are. Five large, strong muscles give the thighs of birds of this order special prominence, and influence the scientific classification. Shy, suspicious skulkers, more fleet of foot than of wing, these birds escape danger by running and hiding rather than by flying.

Cranes

(Family *Gruidæ*)

The cranes, as a family, are birds of largest size, seventeen vertebræ being the usual number in their long necks, and their stilt-like legs elevate their compactly feathered bodies to a conspicuous height. Usually the head is partly bare, or covered with hairlike feathers. Bill is long, straight, slender, and strong. Plumage either white or gray. Solitary wanderers over the

Marsh Birds

plains and marshes, the cranes associate in flocks only at the migrations, although sometimes not averse to feeding in company with other birds, like the geese, for example, as suspicious as they. Field mice, snakes, lizards, frogs, berries, and cereals, all are swallowed by these rapacious feeders. Their voice is harsh, croaking, and resonant, and is frequently heard at night.

Sandhill, or Brown Crane.
Whooping Crane.

Rails, Gallinules, Coots
(Family Rallidæ)

The exceedingly shy, skulking rails, or marsh hens, spend their lives hidden among the sedges of marshes, where they run very lightly over the oozy ground, picking up their food from the surface rather than treading it out of the mud with their long toes. Like the gallinules, they associate with their kin where food is abundant rather than from pure sociability. In spite of their short, rounded wings, they cover immense distances in their migrations; but when flushed in the marshes, where they might remain unsuspected did not their voices betray them, they rise a few feet above the sedges, and, dragging their legs after them, quickly drop down among the grasses they are ever loth to leave. All manner of absurd fables about the rails being blown in from sea, and not hatched from eggs, and certain alleged mysteries of their nests, that human eye, it is said, has never looked upon, are palmed off upon the credulous, not only by the superstitious darkies in southern marshes, but by white people of intelligence, also. Rails are birds of medium or small size; their plumage differs little in the sexes or with age or season; the body is compressed to a point in front, but broad and blunt behind, this wedge shaped figure enabling the bird to squeeze through the mazes of aquatic undergrowth where it finds its constant home. "As thin as a rail" is a truly significant term. Gallinules and coots have a bare, horny plate on the forehead; some of the former are superbly colored. They keep more to the muddy shores of lagoons and ponds and less hidden among the sedges than the rails. Graceful walkers, they are good swimmers also, though their toes lack the lobes that enable the coots to pass much of their lives on the water. Coots live in flocks.

    Clapper Rail.
    King Rail, or Marsh Hen.
    Virginia Rail.
    Sora, or Carolina Rail.
    Yellow Rail.
    Little Black Rail.
    Common, or Florida Gallinule.
    Purple Gallinule.
    American Coot, or Mud Hen.

CRANES
(Family Gruidæ)

Sandhill Crane
(Grus mexicana)

Called also: BROWN CRANE

Length—40 to 48 inches.

Male and Female—Entire plumage leaden gray, more brownish on the back and wings. Upper half of head has dull reddish, warty skin covered with short, black, hairy feathers. Long, acute bill. Very long, stilt-like, dark legs, the tarsus alone being 10 inches long. Tail coverts plumed. Immature birds have heads feathered and more rusty brown in their plumage.

Range—Most abundant in the interior, on the Pacific slope, and the southwest; nests from the Gulf states northward through the Mississippi valley to Manitoba; winters in the Gulf states and Mexico.

Season—Summer resident only north of Florida, Louisiana, and Texas.

Many people confuse this bird with the great blue heron, that is more often called by the crane's name than its own; but beyond a certain resemblance of long legs and necks, these two birds have little or nothing in common.

Immediately on their arrival in the spring the cranes go through clownish performances, as if they were trying to be awkward for the sake of being ridiculous; far from their real intention, however, for it is by these antics that mates are wooed and won. They bow and leap "high in the air," says Colonel Goss, "hopping, skipping and circling about with drooping wings and croaking whoop, an almost indescribable dance and din in which the females (an exception to the rule) join, all working themselves up into a fever of excitement equaled only by an Indian war dance, and like the same, it stops only when the last one is exhausted:"

—strange performances indeed for birds preëminently pompous and circumspect! Certain of the owls and plovers and the flicker also go through laughable antics to win their coy brides, but such boldness of wooing by the female cranes presages the arrival of a "coming woman" among birds, still more nearly approached by the female phalarope, that, without encouragement, does all the wooing.

One may more easily hope to find a weasel asleep than to steal upon a crane unawares. Before settling down to a feeding ground, it will describe great spirals in the air to reconnoitre, the ponderous body moving with slow wing beats, while the keen eyes scrutinize every inch of the region lest danger lurk in ambush. *Grrrrrrrrrrroo*, a harsh, penetrating tremolo calls out to learn if the coast is clear, and *grrrrrrrrrooo* come back the raucous cries from sentinels far and near. Hidden in the grasses, cramped, motionless, breathless, one may be finally rewarded by the alighting of the great stately bird that finally comes drifting downward and stalks over the meadow, alert and suspicious. Not a sound escapes its sharp ears, nor a skulking mouse its even sharper eyes. It will thrust its beak unopened through its prey, whether it is a fish, frog, mouse, or reptile. This terrible weapon makes cowards of the crane's foes, small and large, yet it is the bearer of the spear that is the greatest coward of all.

In addition to animal food, cranes eat quantities of cereals, and when vegetable fed, as they are apt to be in autumn, sportsmen hunt them eagerly, but not too successfully, for no other game bird, unless it is the whooping crane or the wild turkey, so taxes their skill. It is impossible to steal upon them on the open prairie; and in the grass-grown sloughs approach is hardly less difficult. After each bending of the long neck, up rises the head for another reconnoitre. If any unusual sight come within range, the bird stands motionless and tense; then convinced of real danger, "he bends his muscular thighs, spreads his ample wings and springs heavily into the air, croaking dismally in warning to all his kind within the far-reaching sound of his voice," to quote Dr. Coues. In spite of its heavy body the crane rises with slow circlings to a great height until, large as it is, it becomes a mere speck against the clouds. The long neck and stilt-like legs are stretched out on a line with its body, in the attitude made so familiar by the Japanese decorators of our screens and fans. Dur-

Cranes

ing the migrations a flock proceeds single file under the leadership of a wary and hoarse-voiced veteran, whose orders, implicitly followed by each, must first be repeated down the line that winds across the sky like a great serpent.

The Whooping, or White Crane *(Grus americana)*, the largest bird we have, measuring as it does over four feet in length, rarely comes east of the Mississippi, although its migrations extend from South America to the Arctic Circle. Apparently the habits of the two cranes are almost identical, and it is even claimed by some that one alleged third species, the little brown crane, is simply an immature whooper, in which case every feather it owns must be shed before it appears in the glistening white plumage of its parents. Both the whooping and sandhill cranes build nests of roots, rushes, and weed-stalks in some marshy place, and the two eggs of each, which are four inches long, are olive gray, indistinctly spotted and blotched with cinnamon brown.

RAILS, GALLINULES, COOTS
(Family Rallidæ)

Clapper Rail
(Rallus longirostris crepitans)

Called also: MARSH, OR MUD HEN; BIG RAIL; SALT-WATER MEADOW HEN

Length—14 to 16 inches.

Male and Female—Upper parts pale olive varied with gray, each feather having a wide gray margin ; more grayish brown on wings and tail, and cinnamon brown on wing coverts. Line above eye and the throat white, merging into the grayish buff neck and breast ; sides and underneath brownish gray barred with white. Body much **compressed**. Bill longer than head, and yellowish brown, **the same color as legs.** Young fledgelings black.

Range—Atlantic and Gulf coasts of United States, **nesting from** Connecticut southward, and resident south of the **Potomac.**

Season—April to October, north of Washington.

Salt marshes, mangrove swamps, and grassy fields along the seacoast contain more of these little gray skulkers than the keenest eye suspects; and were it not for their incessant chattering, who would ever know they had come up from the south to spend the summer? At the nesting season there can be no noisier birds anywhere than these; the marshes echo with their "long, rolling cry," that is taken up and repeated by each member of the community, until the chorus attracts every gunner to the place. Immense numbers of the compressed, thin bodies, that often measure no more than an inch and a quarter through the breast, find their way to the city markets from the New Jersey salt meadows, after they have taken on a little fat in the wild oat fields. "As thin as a rail" is a suggestive saying, indeed, to the cook who has picked one.

Rails, Gallinules, Coots

To get a good look at these birds in their grassy retreats is no easy matter. Row a scow over the submerged grass at high tide as far as it will go, listen for the skulking clatterers, and if near by, plunge from the bow into the muddy meadow, and you may have the good fortune to flush a bird or two that rises fluttering just above the sedges, flies a few yards trailing its legs behind it, and drops into the grasses again before you can press the button of your camera. A rarer sight still is to see a clapper rail running, with head tilted downward and tail upward, in a ludicrous gait, threading in and out of the grassy maze. Standing on one leg, with the toes of the other foot curled in, is a favorite posture; or one may be detected climbing up the reeds to pick off the seeds at the top, clasping the stem with the help of its low, short, hind toes. A rail's feet are wide spread because of long toes in front, that prevent the bird from sinking into the mud and scum it so lightly runs over. It can swim fairly well, but not fast. As might be expected in birds so shy, these become more active toward dusk, their favorite feeding hour, and certainly more noisy.

Not even to nest will a clapper rail go much beyond tide water. From six to twelve cream white eggs spotted with reddish brown are laid in a rude platform of reeds and finer grasses on the ground, where they must always be damp if not wet; yet who ever finds a mother rail keeping the eggs warm?

. . . .

The King Rail, the Red Breasted Rail, or Fresh Water Marsh Hen (*Rallus elegans*) differs from its more abundant salt water prototype chiefly in being larger and more brightly colored, and possessing a more musical voice. Olive brown, varied with black above; rich chestnut on the wing coverts; reddish cinnamon on breast that fades to white on the throat; sides and underneath dusky, barred with white, are features to be noted in distinguishing it from the grayish clapper. A marsh overgrown with sedges and drained by a sluggish fresh water stream makes the ideal feeding and nesting ground of the king rail from the southern and middle states northward to Ontario. In habits these two rails are closely related. Mr. Frank Chapman describes the king rail's call as "a loud, startling *bup, bup, bup, bup, bup*, uttered with increasing rapidity until the syllables were barely distinguishable, then ending somewhat as it began. The whole performance occupied about five seconds." Of all impossible clews to the identification of a bird, that of its notes as written down differently in every book you pick up is the most hopeless to the novice without field practice. Nearly all the rails have a sort of tree toad rattle in addition to some other notes, which in the king rail's case have a metallic, ringing quality, and that are perhaps most intelligibly written "*ke-link-kink; kink-kink-kink.*"

Virginia Rail

(Rallus virginianus)

Called also: LESSER CLAPPER RAIL; LITTLE RED RAIL; FRESH WATER MUD HEN

Length—8.50 to 10 inches.

Male and Female—Like small king rails; streaked with dark brown and yellowish olive above; reddish chestnut wing coverts; plain brown on top of head and back of neck; a white eyebrow; throat white; breast and sides bright rufous; flanks, wing linings, and under tail coverts broadly barred with dark brown and white; eyes red.

Range—From British Provinces to Guatemala and Cuba; nests from New York, Ohio, and Illinois northward; winters from near the southern limit of its nesting range southward.

Season—Summer resident, April to October, north of Washington.

When the original grant of Queen Elizabeth included nearly all the territory east of the Mississippi that the Massachusetts Bay Colony did not take in, the Virginia rail's name would have been more appropriate than it is to-day; for it is by no means a local bird, as its name might imply, and neither on the coast nor in the interior, north and south, is it rare. Short of wing, with a feeble, fluttering flight when flushed from the marsh, into which it quickly drops again, as if incapable of going farther, this small land lover can nevertheless migrate immense distances. One straggler from a flock going southward recently fell exhausted on the deck of a vessel off the Long Island coast nearly a hundred miles at sea. The ornithologist must frequently smile at the mysteries and superstitions associated with the nesting and migrating habits of this and other rails by the unintelligent.

Doubtless there are many more of all species of rails in the United States than even one who scoured the marshes would suppose. It is only at high tide along the coast that a boat may enter their marshy retreats far enough to flush any birds. The rest, secure in the tall sedges, run in and out of the tall grass on well beaten paths and through aisles of their own making without giving a hint as to their whereabouts. This bird, like the king rail, is frequently called a fresh water, marsh, or mud hen; not because it eschews salt water, but because, even near the sea, it is apt to find out those spots in the bay where fresh water springs bubble up rather than the brackish. Only the bobolinks and red-winged blackbirds, feeding with them on wild oats or rice, the swamp sparrows, marsh wrens, and other companions of the morass, know how many rails are hidden among the bulrushes, sedges, and bushes.

During May, when a nest of grasses is built on the ground, in a tussock that screens from six to twelve pale buff, brown spotted eggs; and in June, when a brood of downy black chicks comes out of the shell, the penetrating voice of the Virginia rail incessantly calls out *cut, cutta-cutta-cutta* to his mate. "When heard at a distance of only a few yards," says Brewster, "it has a vibrating, almost unearthly quality, and seems to issue from the ground directly beneath one's feet. The female, when anxious about her eggs or young, calls *ki-ki-ki* in low tones, and *kiu* much like a flicker. The young of both sexes in autumn give, when startled, a short, explosive *kep* or *kik*, closely similar to that of the Carolina rail." Still another sound is a succession of pig-like grunts, made early in the morning, late in the afternoon, or in cloudy weather. Confusing as are the notes of the different rails, they must be learned if one is to know the shy skulkers, that, unlike a good child, are so much more often heard than seen.

Sora

(Porzana carolina)

Called also: CAROLINA RAIL, OR CRAKE; COMMON RAIL; "ORTOLAN;" SOREE; MUD HEN

Length—8 to 9.50 inches.

Male and Female—"Above, olive brown varied with black and gray; front of head, stripe on crown, and line on throat, black; side of head and breast ashy gray or slate; sides of

SORA RAIL.
⅔ Life size.

breast spotted with white; flanks barred slate and white; belly white." (Nuttall.) Bill stout and short (.75 of an inch long). Immature birds have brown breast, no black on head, and a white throat.

Range—Temperate North America; more abundant on the Atlantic than the Pacific slope. Nests from Kansas, Illinois, and New York northward to Hudson Bay; winters from our southern states to West Indies and northern South America.

Season—Common summer resident at the north; winter resident south of North Carolina; sometimes in sheltered marshes farther north.

Where flocks of bobolinks (transformed by a heavy moult into the streaked brown reed birds of the south) congregate to feed upon the wild rice or oats in early autumn, sportsmen bag the soras also by tens of thousands annually, both of these misnamed "ortolans" coming into market in September and October, by which time the sora's pitifully small, thin body has acquired the only fat it ever boasts. "As thin as a rail" at every other season, however, is a most significant expression, yet many people think it is a fence rail that the adage refers to.

The strongly compressed heads and bodies of all the rail tribe, enabling these birds to thread the maze of aisles among the sedges without causing a blade to quiver and tell the tale of their whereabouts, is almost ludicrous when exposed to view—a rare sight. After one has punted a skiff over the partly submerged grass of their retreats and has waited silent and motionless for endless moments, a dingy little brown, black, and gray bird may walk gingerly out of the reeds, placing one long foot timidly before the other, curling the toes of each foot as it is raised, while with head thrust forward and downward, and with the elevation of the rear end of the body emphasized by the pointed tail that jerks nervously at every step taken, an incarnation of fear moves before you. One old shooter declares he has seen rails swoon and go into fits from fright.

Food gathered from the surface of the ground is picked off with sharp pecks, but all the rails run up the rushes also, clinging with the help of their hind toes to the swaying stem within reach of the grain hanging in tassels at the top. The long front toes, flattened but scarcely lobed, enable them to swim across a ditch or inlet, and all the rails are good divers. Rather than expose themselves as a target for the gunner, they will cling to submerged stalks, with their bills only above water, and allow a skiff to pass over them, without stirring. When thoroughly frightened by the dogs' constant flushing, and the shooting of their masters in the marsh, or, more particularly, when wounded, many never rise again.

It is always the sportsman's hope to flush the rails, whose strong legs and skulking habits sufficiently protect them in the sedges, but whose slow, short flight keeps them within range of the veriest tyro. The 'prentice hand is tried on rails. Trailing their legs after them, and feebly fluttering their wings as they rise just above the tops of the rushes, they soon drop down into them again as if exhausted; yet these are the very birds that migrate from the West Indies to Hudson Bay. Their flight is by no means so feeble as it appears. Darky "pushers" enfold the goings and comings, the nesting and incubation of the rails, with all manner of absurd superstitions.

Were it not for the incessant squeaking, "like young puppies," that is kept up in the haunts of soras, especially at dusk, morning or evening, or at the nesting season, or when startled by a sudden noise, we should never suspect there were birds living in the marshes. Pushers in the reedy lakes of Illinois and Michigan, and along the low shores of the James and other quiet rivers, sweetly whistle and call *ker-wee, ker-wee, peep, peep*, and *kuk, 'kuk, 'huk, k, 'k,'k, 'kuk*, until scores of throats reply, and slaughter soon commences. What little tender flesh there is on the rails' poor bodies, rather flavorless and sapid at the best, is filled with shot for the gourmands to grit their teeth against. As Mrs. Wright says of the bobolinks, so it may be said of the broiled or skewered soras, that they only serve "to lengthen some weary dinner where a collection of animal and vegetable bric-à-brac takes the place of satisfactory nourishment."

In the sedges that shelter and feed them, the rails also build their matted grassy nest, never far from the water, and indeed often lifted into a tussock of grasses washed by it. The eggs, more drab than buff, but spotted and marked with reddish brown like the Virginia rail's, may number as many as fifteen; and the glossy black chicks run about on strong legs, but with the creeping timidity of mice, from the hour of hatching.

. . . .

The Yellow, or New York, or Yellow-breasted Rail (*Porzana*

noveboracensis), an even more skulking, timid species than the sora, has a reputation for rarity that doubtless the blackbirds, bobolinks, and marsh wrens, which alone can penetrate into the mysteries of the sedges, would express differently were they able to retail secrets. This small rail, that measures only seven inches in length, has more wisdom than its larger kin, and refuses to be flushed except in extreme cases, for the gunners to hit during its feeble, fluttering flight. Dogs must be sent into the marshes after the panic stricken birds running through aisles of grasses until about to be overtaken, when they escape by rising from the frying pan of the dogs' jaws only to fall into the fire of shot from the rifles. Ordinarily they keep so closely concealed among the grasses, that were it not for their croaking call, suggesting the voice of the tree toad, no one would suspect their presence. All rails are more or less nocturnal in their habits, and the yellow-breasted species, more full of fears than any, rarely lifts up its voice, that Nuttall described as an "abrupt and cackling cry *'krèk, 'krèk, 'krèk, krèk, kuk, k 'uk,*" after daylight or before sunset. The description of the sora's habits, which are almost identical with this rail's, should be read to avoid repetition. In plumage, however, these two birds are quite different, the yellow-breasted rail having black upper parts streaked with brownish yellow and marked with white bars, the buff of the breast growing paler underneath, the dusky flanks barred with white, and the under coverts varied with black, white, and rufous. Its wing linings are white, but these the bird takes good care not to show.

. . . .

The Little Black Rail, or Crake (*Porzana jamaicensis*), the smallest of the family, exhibits all the family shyness and fear, which, taken with its obscure coloration and its extreme unwillingness to rise on the wing, keep it almost unknown, although its range extends from Massachusetts, Illinois, and Oregon to Louisiana, the West Indies, and Central America. As its name implies, it is common in Jamaica. Mr. Marsh of that island writes its call "*chi-chi-cro-croo-croo*, several times repeated in sharp high notes so as to be audible to a considerable distance." Guided by this call, one may count oneself rarely fortunate to discover the little mouse-like bird that makes it, running swiftly in and out of the sedges. Its head, breast, and under parts are slate color; its fore back and nape are rich brown; its lower back, wings, and tail are brownish black spotted with white, and the flanks and dusky under parts are barred with white.

Common Gallinule
(Gallinula galeata)

Called also: FLORIDA GALLINULE; WATER HEN; RED-BILLED MUD HEN; BLUE RAIL

Length—12 to 14 inches.

Male and Female—A bare, bright red shield on forehead, same color as bill; plumage uniform dark bluish or grayish black, darkest on head and neck; washed with olive brown on back and shoulders, and fading to whitish underneath; flanks conspicuously streaked with white; space under tail white; legs greenish yellow, reddish at joint.

Range—Temperate and tropical America, nesting from Ontario and New England to Brazil and Chili, and wintering from our southern states southward.

Season—Summer resident or transient summer visitor, from May to October, north of the southern states.

There is a popular impression, for which the early ornithologists are doubtless responsible, that all gallinules are birds of the tropics; but this so-called Florida species crosses the Canadian borders in no small numbers every summer, and nests are also constantly reported in our northern and middle states. The truth probably is that the range of the Florida gallinule has not extended, but that within the last half century a hundred bird students scour our woods, meadows, and marshes for every enthusiast that tramped over them fifty years ago; and we are just becoming thoroughly acquainted with many of our birds when the gunners, milliners, cats, and other fatal accompaniments of a civilization that in many respects is still barbaric, threaten to exterminate the sadly decreased numbers left us to enjoy.

Gallinules, although wild, shy, and timid creatures, or they would be no kin of the rails, wade more than they and swim expertly. It is amusing to watch their heads bob in rhythm with their feet as they rest lightly on the water. In brackish pools rather than salt ones, and preferably around fresh water

PURPLE GALLINULE.

lakes and meadow brooks, they keep well concealed among the sedges while the sun is high or when danger threatens, coming boldly out to feed on the mud flats at dusk, or when they think themselves unobserved. Apparently they tolerate other gallinules' society only if they must. Quarrels arising from jealousies over an infringement of territorial rights frequently occur.

A gallinule strides from its grassy screen with grace and elegance, curling its toes when it lifts its large foot, as if it had taken a course of Delsarte exercises. Wading into the shallow pool, still curling its long toes before plunging its foot downward, and tipping its tail at every step, showing the white feathers below it, the bird strides along, close to the shore, stopping from time to time to nip the grasses and seeds on the bank, or to secure some bit of animal food on the muddy bottom of the water. Snails and plantains are favorite morsels. When lily pads or other flat leaved plants appear on its path, the gallinule runs lightly over them, upheld partly by its long toes and partly by its fluttering wings. Dr. Abbott tells of seeing a gallinule in his favorite New Jersey creek that went through the unusual (?) performance of throwing back its head until the occiput rested on its shoulders, and at the same moment the wings were lifted lightly as if the bird intended to fly.

But flying is an art this terrestrial wader practices rarely. It depends sometimes upon swimming and diving, but almost always on running, to escape danger, many men of science claiming that a large part of its migrating also is done a-foot. As the family parties escape under cover of darkness, and steal away as silently as the Arabs, who knows positively how they travel? A gallinule, equally with a barnyard chicken, appears ridiculous and out of its element in the air as it labors along a few paces, dragging its legs after it, and drops awkwardly to the ground.

The similarity to a chicken does not end with flight. In appearance, as in habits, and particularly in voice, the water hens and hens of the poultry yard have much in common. A single pair in a swamp keep up clatter enough for a yard full of fowls, "now loud and terror stricken like a hen whose head is just going to be cut off," as a friend of Bradford Torrey's expressed it; "then soft and full of content, as if the aforesaid hen had laid an egg ten minutes before and were still felicitating herself upon the achievement." When both the Florida and the purple gallinules build their nests, they very often simply bend down the tops of grasses to form a platform, then place a rude, grassy cradle on it; or the nest may be moored to the stems of the rushes, or to a bush, where the incoming tide raises it, but cannot loosen its anchors. But usually drier sites are chosen.

. . . .

The Purple Gallinule (*Ionornis martinica*), a common bird in the southern states, nests so far north as southern Illinois and Carolina, and occasionally strays northward to New England and Wisconsin. In the Gulf states it is usually found in the same marsh with the Florida gallinule, eating the same food, nesting in the same manner, cackling like a chicken, in fact sharing nearly all its cousin's habits, its gorgeous plumage alone giving it distinction.

American Coot
(*Fulica americana*)

Called also: WHITE-BILLED COOT; CINEROUS COOT; MUD HEN; CROW DUCK; BLUE PETER; MOOR HEN; MEADOW HEN

Length—14 to 16 inches.

Male and Female—General color slate; very dark on head and neck, lighter on under parts; edge of wing, tips of secondaries, and space below tail, white. Bill ivory white; two brownish spots near tip, the same shade as the horny plate on front of head, which is a characteristic mark of both gallinules and coots. Legs and feet pale green, the latter with scalloped lobes.

Range—North America at large, from Greenland and Alaska to the West Indies and Central America; nesting throughout range, but more rarely on Atlantic coast.

Season—Resident in the south; chiefly a spring and autumn migrant at the north, April, May; September to November.

More aquatic than any of its kin, the coot delights in the swimming and diving feats of a grebe, and appears to be the connecting link between the swimmers, with whom it was formerly classed, owing to its lobed toes. What these toes lack in width is amply made up in length, the fact that makes the bird so expert in the water and correspondingly awkward when it runs over the land, where, however, it spends very little time. It is the horny frontal plate, taken with the general resemblance in structure to the gallinules, that places the coot in their class.

A lake or quiet river surrounded by large marshy tracts where sluggish streams meander, bringing down into deeper water wild grain and seeds, the larvæ of insects, fish spawn, snails, worms, and vegetable matter, makes the ideal home of this duck-like bird. "I come from the haunts of coot and hern," the song of Tennyson's brook, calls up a picture of the home that needs no enlarging. The coot dives for food to great depths, sometimes sinking grebe fashion, and disappearing to parts unknown by a long swim under water with the help of both wings and feet. Swimming on the surface, the bird has a funny habit of bobbing its head in unison with the strokes given in the stern by its twin screws.

A large amount of gravel seems necessary to help digest the quantity of grain swallowed, and for this a flock of coots must sometimes leave the muddy region of the lake. Rising from the surface, they flutter just above it, pattering along for a distance, their distended feet striking the water constantly, until sufficient momentum is gained to spring into the air and trust to wing power alone. This pattering noise and splashing, often heard when the coots cannot be seen for the tall sedges that screen them, is characteristic of several of the ducks also, and suggests the notion that the trick may have been learned from them; for in southern waters, at least, coots and ducks often resort to the same lakes ;—that is, when the latter refuse to be driven off. . At no time of the year silent birds, often incessant chatterers, it is during the nesting season that the coots break out into shrill, high-pitched, noisy cacklings, which the slightest disturbance calls forth. Jealous, unwilling to permit alien swimmers in their neighborhood, sociable, but without any great love of kin or kind to mellow their dispositions or their voices, they make their neighborhood lively. But coots are shy of men, albeit the young and old alike have flesh no one not starving could eat; and they usually live in some inaccessible pond or swamp, especially at the nesting season. As night approaches, they lose much of the timidity which keeps them concealed and silent the greater part of the day.

Rails, Gallinules, Coots

In May a nest has been built by first trampling down the rushes and weed stalks, then more of the same material is used for an exterior and finer grasses for a lining of the crib which toward the end of the month contains from eight to fifteen yellowish white eggs sprinkled over with brownish spots, chiefly around the larger end. Let no other bird dare show its head in the immediate neighborhood of a pair of nesting coots. They will tolerate no neighbors then, gregarious as they are at other seasons. After three weeks of close confinement the mother bird leads her large brood to water, where the chicks swim and dive almost from the beginning, although keeping close enough to their patient teacher to hide under her wings on the first shrill alarm cry from the father, ever on guard. Hawks from above and pickerel and turtles from below find no fault, as men do, with the flavor of young coots. But soon the fledgelings become quite independent, leaving the parents free to devote their attention to another brood. Usually the flock of migrating coots that we see in autumn is only a large family party.

AMERICAN COOT.
½ Life-size.

SHORE BIRDS

Phalaropes
Avocets
Stilts
Snipe
Sandpipers
Plovers
Turnstone
Oyster Catcher

SHORE BIRDS

PHALAROPES, SNIPE, SANDPIPERS, PLOVERS,
SURF BIRDS, ETC.

(Order Limicolæ)

Birds of the open field, marshy bogs and thickets, or shores close by the water's edge, finding their food on the surface of the ground, in the mud, or among the shallows of the beach, averaging smaller than birds of any other group included in this book, they usually have long slender legs for wading, and long slender bills for probing the mud after food, which increase their apparent size. Unlike the compressed figures of rails and their allies, the bodies of these birds are depressed or well rounded; their wings are long and pointed; their tails, which are short, are very full feathered. As compared with the large footed herons, rails, and gallinules, these birds have short toes, the hinder one very short, elevated, or absent; but certain species find their toes long enough to tread out worms and small shell fish from the mud flats, and some, partly webbed, are well adapted for swimming. The nests of birds of this very large order are mere depressions in the ground, not always lined with grass, and their young, fully clothed with down when hatched, are able to run about immediately.

Phalaropes

(Family Phalaropodidæ)

A small, select family of three, two of whose species keep so far out at sea during their migrations from the Arctic regions to the south that we rarely see them, Wilson's phalarope, alone, being anywhere a common bird in the United States. Strangely enough, it is far more abundant in the interior than on the coast.

Shore Birds

The bodies of these small sea snipe, as they are often called, are depressed, covered with thick plumage to resist water that they spend much time upon, for their feet are furnished with narrow lobes that enable them to swim well. They are smaller than the robin. The curious characteristic of this family is that it contains the most advanced female among all the feathered tribes; this strong minded creature wearing the gay colors, doing the wooing, and gayly disporting herself, while the male incubates the eggs and attends to nursery drudgeries.

Wilson's Phalarope
Northern Phalarope

Avocets and Stilts

(Family Recurvirostridæ)

Usually one sees a small flock of these waders, very long of legs, slender and depressed of body, and with a long, sharp bill, curved upward like an upholsterer's needle or a shoemaker's awl. This bill, which is of extreme sensitiveness, probes the mud in the shallows where the birds wade about for food. Sometimes called wading snipe, they swim, when necessary, as easily and gracefully as they walk. Their plumage may differ with the season, but the sexes and young are alike.

American Avocet
Black-necked Stilt

Snipe, Sandpipers, etc.

(Family Scolopacidæ)

Generally the sensitive bill is long and straight, often several times longer than the head, and frequently curved slightly upward or downward. With this tool these birds probe the sand or mud for food, feeling for what they want, and using the bill also as a forceps. Often the upper prong may be bent at will for hooking the earthworms out. Birds of this numerous family have four toes instead of three; but, in most instances, the structure is very like that of the plovers. Plumage, which is plain colored, varies with the season, but little with the sexes or with age. Usually the female is the larger. These birds average small, the least sandpiper being the smallest of our water fowl. With few exceptions they keep near the water's edge or wherever the ground is soft enough to be easily probed, whether by the sea and rivers or in inland bogs, moist meadows, and thickets. Exclusive when nesting, but not often solitary at other seasons, they are generally gregarious, strongly attached to their companions, and migrate in large flocks. "The voice is a mellow pipe, a sharp bleat, or a harsh scream, according to the species," says Dr. Coues. "Few birds surpass the snipe in sapid quality of flesh, and many kinds rank high in the estimation of the sportsman and epicure."

Woodcock
Wilson's or Jack Snipe
Dowitcher
Long-billed Dowitcher
Stilt Sandpiper
Knot or Robin Snipe
Pectoral Sandpiper
White Rumped Sandpiper
Baird's Sandpiper
Least Sandpiper
Red-backed Sandpiper
Semipalmated Sandpiper
Western Semipalmated Sandpiper
Sanderling or Surf Snipe
Marbled Godwit or Brown Marlin
Greater Yellowlegs
Yellowlegs
Solitary Sandpiper or Tatler
Willet
Bartramian Sandpiper or Upland "Plover"
Buff-breasted Sandpiper
Spotted Sandpiper
Long-billed Curlew
Jack Curlew
Eskimo Curlew or Doe Bird

Shore Birds

Plovers
(Family Charadriidæ)

Resembling the snipe in structure, plovers may be distinguished by their moderate or small size, averaging that of the thrush, by their short bills (not longer and generally shorter than the head), which are shaped somewhat like a pigeon's; by their three toes—not an infallible guide, however, since our black-breasted species and two others have four toes;—in having rounded scales on the tarsi; by their plump bodies, short, thick necks, long wings, reaching to the tip of the tail or beyond, and, in some instances, by spurs on the wings. In habits, too, there is a similarity to the preceding group; but the plovers pick their food, which is largely of an animal nature, from the surface of the ground, instead of probing for it, as their shorter bills indicate. They also more frequently visit dry fields and uplands. Rapid runners and fliers, mellow whistlers, gregarious, except at the nesting season, and not shy, plovers are among the best known of our common birds.

    Black-breasted Plover or Beetle-head
    Golden Plover
    Kildeer
    Semipalmated or Ring-necked Plover
    Piping Plover
    Belted Piping Plover
    Wilson's Plover

Surf Birds and Turnstones
(Family Aphrizidæ)

One member only of this maritime family of four species visits the outer bars and beaches of our sea coast, to turn over shells and pebbles looking for the small animal life it preys upon. Its head and bill resemble a plover's; its wings are long and sharply pointed for sea roaming.

    Turnstone or Calico-back

Oyster Catchers
(Family Hæmatopodidæ)

The brightly colored bill, twice as long as the head, compressed like a knife blade toward the end, is the chief distinguishing mark of birds of this small family. This tool is used to pry open the shells of mussels, oysters, clams, and other shellfish; hence Dr. Coues suggests oyster-opener as a better name for these birds, since oysters don't run fast! Rather large birds, dark colored and white in masses; the plumage of the sexes similar; the legs stout and rough; no hind toe; the wings long and pointed for long sea flights.

American Oyster Catcher

PHALAROPES
(Family Phalaropodidæ)

Wilson's Phalarope
(Phalaropus tricolor)

Called also: SEA SNIPE; SWIMMING SANDPIPER; LOBE-FOOTED HOLOPODE; SEA GOOSE.

Length—8.25 to 9 inches. Smaller than a robin; female the larger.

Female: In summer—"Top of the head and middle of the back pearl gray, nape white; a black streak passes through the eye to the side of the neck, and, changing to rufous chestnut, continues down the sides of the back and on the scapulars; neck and upper breast washed with pale, brownish rufous; rest of the under parts and upper tail coverts, white.

Male: In summer—Upper parts fuscous brown, bordered with grayish brown; upper tail coverts, nape, and a line over the eye white or whitish; sides of the neck and breast washed with rufous; rest of the under parts white.

Adults: In winter—Upper parts gray, margined with white; upper tail coverts white; wings fuscous, their coverts margined with buffy; under parts white."—(Chapman.)

Range—Temperate North America, most abundant in the interior; nesting from northern Illinois and Utah northward, and wintering southward to Brazil and Patagonia.

Season—Chiefly a migrant in the United States; more rarely a summer resident.

Without the help of the woman's college, club, or bicycle, the female phalarope has emancipated herself from most of the bondages of her sex, showing a fine scorn for its conventional proprieties. It is she who, wearing the handsome feathers and boasting a larger size than the male,—although neither bird is so large as a robin,—undertakes to woo her coy sweetheart by bold advances. Possibly a brazen rival adds to his miseries. The at first reluctant lover may run away, but, quickly overtaken, he soon falls a victim to the wiles of the most persistent wooer, to continue the most hen-pecked of mates ever after.

On him fall all the domestic drudgeries, except the laying of the eggs—the one feminine accomplishment of his almost unsexed boss. He chooses the site for their nursery in a tuft of grass in a wet meadow or soft earth, usually near water; and, having scratched a slight depression in the soil and lined it with grass, she actually condescends to lay three or four cream colored eggs, heavily blotched with chocolate brown, about the first of June. Sometimes a second and smaller set of eggs is found late in the season. Many male birds, as we all know, relieve their brooding mates, but is there another instance where the male does all the incubating, while the female enjoys life at ease? What must a totally enslaved mother duck think of such emancipation? And what compassion must not a dandified, care-free drake feel for the male phalarope confined on the eggs day after day, and scarcely permitted twenty minutes for refreshments?

To secure their food, phalaropes run along the marshes and beaches exactly like sandpipers, picking up snails and other small animal forms, and nodding their heads as they go; or wading knee deep into the ponds, thrust them below the shallow water. "Swimming Sandpipers" they certainly are, though they swim rarely, never for long at a time, or in deep water. Every movement, whether afloat or ashore, is full of daintiness and grace. In flight they sometimes cover short distances in a zigzag, as if uncertain of their direction; but once launched on a long migration, they fly with directness and power.

. . . .

The Northern Phalarope *(Phalaropus lobatus)*, a very small, slaty gray, chestnut red, buff and white bird, the smallest of all the swimmers, passes along the coasts of the United States, from its nesting grounds in the Arctic regions, to winter in the tropics. Great flocks, bedded or swimming in the ocean, are often met by coastwise steamers in spring and from August to November.

AVOCETS AND STILTS

(Family Recurvirostridæ)

American Avocet

(Recurvirostra americana)

Called also: BLUE STOCKING; WHITE SNIPE; SCOOPER.

Length—16 to 20 inches.

Male and Female: In summer—White, changing into cinnamon, on neck and head; shoulders and wings brownish black, except the middle coverts, the tips of the greater ones, and part of the secondaries, which are white. Very long, excessively slender black bill, curved upward. Legs very long and of a dull blue. *In winter:* Similar, but head and neck ashy or pearl gray like the tail.

Range—Temperate North America, nesting from Texas northward to Great Slave Lake, and wintering in Central America and the West Indies. Rare in the eastern United States. Irregularly common in the interior.

Season—Summer resident or spring and autumn migrant.

The avocet, like the skimmer, the sea parrot, and the curlew, possesses one of the most extraordinary bills any bird wears. Slowly swinging it from side to side, as a farmer moves his scythe, the eccentric looking bird wades about in the shallows, feeling on the bottom for food that cannot be seen through the muddy water. Often the entire head and neck must be immersed to probe the mud for some small shell fish and worms that the sensitive, needle-like bill dislodges. A leader usually directs the motions of a small flock that follows him through thick and thin, mud and water; or, if the water suddenly deepens, off swim the birds until their feet strike bottom again, and the mowing motion is resumed, while the sickle bills feel and probe and jerk as the mowers move along deliberately and gracefully. The curlew's tool, the true

AMERICAN AVOCET.
½ Life size.

sickle-bill, curves downward, just the reverse of the avocet's; neither is it used under water.

The avocet is, perhaps, the best swimmer among the waders, owing to its webbed toes. The thick, waterproof plumage of its under parts keeps its body dry. When about to alight it chooses either water or land, indifferently; but it is always especially abundant in or about the alkaline marshes of the interior. Not at all shy of man, it pays little attention to him unless positively pestered, when, springing into the air, and trailing its long legs stiffly behind to balance its outstretched neck, it flaps leisurely away to no great distance, calling back *click, click, click*, a sharp and plaintive cry. A long sail on motionless wings, and a drift downward, brings the bird to the ground again, but tottering at first, as if it took time to regain its equilibrium, just like a stilt. On alighting, it strikes an exquisite pose, lifting its wings till they meet over its back, like the terns and plovers, before folding them away under the feathers on its side.

The nest is a mere depression in the ground, in a tuft of thick grass growing in some marshy place, and it may be lined with fine grasses, though such luxury is not customary. Three or four pale olive or yellowish clay colored eggs, thickly spotted with chocolate brown, are a complement. Near such a spot, the birds become clamorous and excitable, the entire colony resenting any liberty taken by an intruder carrying no more alarming weapon than a field glass. Still, a male avocet, lost in rose colored day dreams as he paces up and down near his nest, like the willet, on sentinel duty, rarely sees anything that is not directly in his way.

Black-necked Stilt

(Himantopus mexicanus)

Called also: LAWYER; LONGSHANKS; TILT; TILDILLO; WHITE SNIPE.

Length—About 15 inches.

Male and Female—Mantle over back and wings black, also line running up back of long neck and spreading over top and sides of head below the eye. Tail grayish; rest of plumage,

Avocets and Stilts

including a spot above and below the eye, white. (Long black wings, folding over white spots on lower back, rump, and upper tail coverts, make the entire upper parts appear black.) Immature birds more brownish above. **Long, straight, slender, black bill.** Excessively long red or pink legs. Beautiful large crimson eyes.

Range—Tropical America, nesting northward from the Gulf states, "locally and rarely" up the Mississippi; rare on the Atlantic coast, though specimens have been taken in Maine and some reach Long Island annually; most abundant in the southwest.

Season—Summer resident or visitor. Permanent resident in Gulf states.

To a query put to an Arkansas farmer as to why this bird should be called the lawyer, immediately came another query: "Ain't you ever noticed its long bill?"

But it is the excessive length of legs that attracts the attention of all except punsters. So slender and stilt-like are they, so teetering and trembling is the bird when it alights, that one's first impulse is to rush forward and help it regain its equilibrium before it falls. Why must the stilt always go through this pretense of feebleness when we know it is a strong steady walker, graceful and alert; or, does it actually lose its balance on alighting?

Wading about, with decided and measured steps, in shallow pools, preferably among the salt and alkaline marshes, where the avocets often keep them company, the stilts pick up, from first one side, then the other, insects and larvæ, small shell fish, worms, fish fry, etc., often plunging both head and neck under water to seize some deep swimmer. Long as their legs are, they will wade up to their breasts to secure a good meal; but, having no webs to their toes, swimming does not come easy, as it does to avocets, nor is it often tried.

Strong fliers, owing to their long wings, which, when folded, reach beyond the tail, the longshanks trail their stiffened legs behind them at a horizontal, after the manner of their tribe, and continually yelp *click, click, click,* as the flock moves leisurely overhead. In the nesting grounds this yelping cry is incessant, however far the intruder keeps from the olive or clay colored eggs or the young chicks that run about as soon as hatched.

SNIPE, SANDPIPERS, ETC.
(Family Scolopacidæ)

Woodcock
(Philohela minor)

Called also: BLIND, WALL-EYED, MUD, BIG HEADED, WOOD, and WHISTLING SNIPE; BOG-SUCKER; NIGHT PECK; BOG BIRD; TIMBER DOODLE; NIGHT "PARTRIDGE"; PEWEE.

Length—10 to 11 inches; Female 11 to 12 inches.

Male and Female—Upper parts varied with gray, brown, black, and buff; an indistinct black line on front of head, another running from bill to eye; back of head black with three buff bars. Under parts reddish buff brown. Eyes large and placed in upper corner of triangular head. Bill long, straight, stout. Short, thick neck and compact, rounded body; wings and legs short.

Range—Eastern North America, from the British Provinces to the Gulf, nesting nearly throughout its range; winters south of Virginia and southern Illinois.

Season—Resident all but the coldest months; a few winter.

The borings of the woodcock in bogs, wet woodlands, and fields—little groups of clean cut holes made by the bird's bill in the soft earth—give the surest clue to the presence of this luscious game bird, that has been tracked by sportsmen and pot hunters alike, from Labrador to the Gulf, by means of these tell-tale marks until the day cannot be far distant when there will be no woodcock left to shoot. Since earthworms are the bird's staple diet, these must be probed for and felt after through the moist earth. Down goes the woodcock's bill, sunk to the nostril; the upper half, being flexible at the tip, draws the worm forth as one might raise a string through the neck of a jar with one's finger. Curiously, the tip of the upper mandible works quite independently of the lower one—a fact only recently discovered by Mr. Gurdon Trumbull. Owing to the position of the eyes, at the back of the head, food must be felt rather than seen; but, so sensitive is the tip of the bill, and so far out of sight are the worms, in any case the eyes serve a better purpose in being placed where they widen the bird's vision and so detect an enemy afar. It is claimed by some that, like the owls, woodcock see best at night. Worms come to the surface after dark, which explains this and many other birds' nocturnal habits.

In the early spring any one who takes an interest in the woodcock, aside from its flavor, will be repaid for one's tramp through the swale, at evening, to see the bird go through a series of aërial antics and attestations of affection to his *innamorata*. Standing with his bill pointing downward and his body inclined forward, he calls out *pink, pink*, as much as to say: "Now look, the performance is about to begin"; then suddenly he springs from the ground, flies around and around in circles, his short stiff wings whistling as he goes, higher, higher, faster, faster, and louder and louder, as he sweeps by overhead in erratic circles, each overlapping the other, until the end of the spiral described must be fully three hundred feet from the ground. Now, uttering a sharp whistle, down he comes, pitching, darting, and finally alighting very near the spot from which he set out. *Pink, pink*, he again calls, to make sure his efforts are not lost upon the object of his affection, and before he can fairly have recovered his breath, off he goes on another series of gyrations accompanied by wing music. Or, he may dance jigs when in the actual presence of the loved one. Cranes, plovers, owls, and flickers, among others, go through clownish performances to win their mates, in some instances the females joining in; but the woodhen, as the proper-nice people say, remains coy and apparently coldly indifferent to the madness of her lover. He will sometimes stand motionless, as if meditating on some new method of winning her, his head drawn in, his bill pressing against his breast. Then, with his short tail raised and outstretched like a grouse's, and with dropped wings trailing beside him, he will strut about with a high step—a comical picture of dignity and importance.

Little time need be taken from the honeymoon to make a nest. This consists of a few dry leaves on the ground in the

AMERICAN WOODCOCK

woods, usually near a stump, where the four buffy eggs, spotted over with reddish brown, are laid, often before the snow has melted, in April. A dry place being chosen for the nesting site, it sometimes becomes necessary to transport the funny little fluffy, long-billed chicks to muddy hunting grounds, and the mother has been detected in the act of flying with one of her brood held between her thighs. But the chicks are by no means helpless, even from the instant they leave the shell. It is a pretty sight to see a little family poking about at twilight for larvæ, worms, and small insects, among the decayed leaves, the fallen logs, and the ferns and skunk cabbages. *Peep, peep,* they call, quite like barnyard chicks.

By the first of August the woodcocks, deserting the low, wet lands, scatter themselves over the country in corn fields, grassy meadows, birch covered hillsides, "alder runs," pine forests, and thick, cool, moist undergrowth, near woods; and now they moult. No whistling of wings can be heard as the birds heavily labor along near the ground, often unable to raise their denuded bodies higher. In September, when the sportsmen make sad havoc in the flocks, already gathering for migration, they are found in the dense thickets of wooded uplands, where a stream flows to keep the ground soft; and in October, when the birds are in prime condition, the spot that contained scores at evening may hold none by morning. The russet colored birds mingle with the russet colored leaves, and, as they lie close, it takes a good dog to find them. The woodcocks migrate silently by night, and an early frost, that stiffens the ground, drives them off suddenly to softer territory southward. Hence the delightful element of uncertainty enters into the hunting of this bird, that is here to-day and gone to-morrow. When flushed, its flight appears to be feeble, as, after a few whistles of its short, stiff wings, and trailing its legs behind it, it quickly drops into cover again, running a little distance on alighting, but the distances covered in migrations prove it to be no unskilled flier.

Wilson's Snipe

(Gallinago delicata)

Called also: "ENGLISH" SNIPE, COMMON SNIPE; JACK SNIPE; AMERICAN SNIPE; SHAD BIRD.

Length—10.50 to 11.50 inches.

Male and Female—Upper parts varied with black, brown, and buff; crown dusky, with buff stripe; throat white; neck and breast buff, streaked with dusky; underneath white, the sides with blackish bars. Outer feather of wings white; wings brownish black, the feathers barred with reddish brown and margined with white. Tail bay and black, the outer feathers barred with black and white; the inner ones black, marked across the end with rufous and tipped with soiled white. Bill about 2.50 inches long and resembling the woodcock's.

Range—North America at large, from Hudson Bay and Alaska, south in winter to central and northern South America and the West Indies. Nests in far north chiefly, rarely in the northern United States.

When the first shad run up our rivers to spawn, and the shad bush opens its feathery white blossoms in the roadside thickets in March, the snipe come back from the south to haunt the open wet places of the lowlands, fresh water marshes, soaked fields, and the sheltered sunny spots in a clearing that are the first to thaw. Only in exceptionally dry seasons do these birds go near salt water marshes. Generally speaking, snipe prefer more open country than woodcock; but plenty of the former have been flushed in bush-grown, springy woods—the woodcock's paradise when the lowlands become flooded. The russet colors and markings of these birds, that so perfectly mimic their surroundings as they lie close, conceal them from all but the sharpest eyes. We may know of their arrival by the clusters of holes in the mud; for both snipe and woodcock have the habit of thrusting their bills into the soft ground up to the nostrils, feeling for worms as they probe with the sensitive tip whose upper half is flexible and capable of hooking the earthworm from its hole. As the snipe's eyes are set far back in its head, it must be guided only by the sense of touch. The larvæ of insects and insects themselves are found by overturning old leaves and decayed wood; but most of this bird's food must be probed for. Martin Luther was not the only one to profit by a Diet of Worms!

While comparatively few nests are built in the United States, most of the love making is done here, and one of the characteristic spring sounds in districts frequented by this snipe is the Æolian whistling of its wings at evening, dawn, or by moonlight, when its wooing is done chiefly in mid air. Lighter and more trim of figure than a woodcock, Wilson's snipe is a better flier, and, rising upward by erratic yet graceful spirals, it attains a height we can only guess at but not see in the dusk; then darting earthward, music thrums and whistles in its wake to charm the ear of the listening sweetheart. It makes "at each descent a low yet penetrating, tremulous sound," says Brewster, "which suggests the winnowing of a domestic pigeon's wings, or, if heard at a distance, the bleating of a goat, and which is thought to be produced by the rushing of the air through the wings of the snipe. . . . Besides this 'drumming' or 'bleating,' as it is called, the snipe, while mating, sometimes make another peculiar sound, a *kuk-kuk-kuk-kuk-kup*, evidently vocal, and occasionally accompanying a slow, labored, and perfectly direct flight, at the end of which the bird alights on a tree or fence post a few minutes."

The flight of a snipe, almost invariably erratic, zig-zag one minute and maybe strong and direct the next, discourages all but the most expert wing shot. Although lying close, and generally flushed in the open, no tyro is quick enough at covering the swift, tortuous flier to bag it. Nervous, excitable, and therefore particularly difficult to hit, poor of flesh and muscular from long travel in the spring migration, nevertheless there are in many states no laws to prevent the killing of these snipe then; and the fact that eggs are already formed in many birds brought to the kitchen has not yet moved the hearts of sportsmen and legislators to action. For the most part, these snipe go north of the United States to lay three or four clay-colored or olive eggs, heavily marked and scratched with chocolate, in a depression in the ground.

When the early frosts of autumn harden the soil at the north, so that the bill can no longer penetrate it, the snipe, migrating by night, again visit us, this time fatter, more lazy, or at any rate less nervous than they were during the mating season. Just as a wet meadow may be full of them some August morning before we are expecting them, so in September the sportsmen go to look for them at dawn where they were the evening before in numbers, to find that they have silently travelled southward during the night. There is always the charm of the unexpected about the snipe's appearance or disappearance. Like the woodcock, it is almost nocturnal in habits, because earthworms come to the surface then. Coming out from under cover, where it has dozed the best part of the day, to feed in the open at twilight of morning or evening, it lies close until flushed, when, springing upward from the grass almost at the sportsman's feet, as if shot out of a spring trap, and startling the novice out of a good aim by its hoarse, rasping *scaip, scaip*, it stands a good chance of escaping, thanks to its swift zigzag course.

Dowitcher

(Macrorhamphus griseus)

Called also: RED-BREASTED SNIPE (summer); QUAIL SNIPE; BROWN JACK; GRAY SNIPE (winter); DOWITCHEE; BROWN BACK; ROBIN SNIPE; DEUTSCHER or GERMAN SNIPE.

Length—9.50 to 10.50 inches.

Male and Female: In summer—Upper parts black, the feathers edged or barred with rusty red, white, and buff; tail and rump white barred with dusky; lower part of back white, conspicuous in flight; under parts rusty red, paler or white below, more or less spotted and barred with dusky. Bill, which is two inches long, is blackish brown. Legs and feet greenish brown. *In winter*—General plumage brownish or ashy gray; lower back white; rump and tail barred with dusky and white; lower parts white, shading into gray on breast.

Range—Eastern North America, nesting within the Arctic Circle and wintering from Florida to the West Indies and Brazil.

Season—Spring and autumn migrant; April, May; August and September.

Compact flocks of gray snipe, as they are called after the summer moult has transformed them, migrating southward along the sea coast in August and September, may be easily

WILSONS SNIPE.

called down by anyone sufficiently familiar with their loud, quivering, querulous whistle to imitate it. Sportsmen also use decoys; but these are gentle, sociable birds, among the last to suspect evil or to take alarm, and need little encouragement to alight beyond the supposed entreaties of a sister flock. They appear to be never in a hurry; the long journey to and from their nesting grounds has frequent halting places; the mellow days of early autumn find them free from care and ready to accept every invitation to enjoy life to the full.

Wheeling about as the imitation of their call reaches them, if they are not perchance flying too high to hear it, down swings the flock, hovering over the mud flats and tracts of low beach exposed at ebb tide. After circling about and seeing none of their kin, they may nevertheless decide to stop and rest awhile and feed in so promising a field. Now they scatter, but never so far that a chattering talk may not be kept up with their companions while they look for snails, seeds of sedges, insects, small mollusks, gravel, and bits of vegetable matter picked off the surface or from the shallow pools in the salt marshes. Sometimes they probe the soft mud, too, for some tiny marine creature that has buried itself there; but not commonly, as the woodcock and Wilson's snipe do. A sand bar will often be so crowded with these sociable little waders that the sportsman picks off a dozen or more birds at a single shot; and so innocent are they that even such a lesson does not prevent their returning to the identical spot after a short flight. It is small wonder they are favorites with shooters.

Skimming over the marshes, swallow fashion, a flock darts about in an erratic, joyous course—now high in air and performing some beautiful evolutions, now close above the sedges—their shrill, quivering whistle, constantly called back and forth, keeping the neighborhood lively. The note can scarcely be distinguished from the whistle of the yellowlegs that these snipe frequently associate with as they do with various sandpipers. When on the wing, the white spot on the lower back, a diagnostic feature, is conspicuous enough to help the novice name the bird.

A number of nests or depressions in the moss or grasses that answered the purpose, have been found near lakes and marshes at the far north by travellers who have brought back to our museums clutches of four drab or fawn colored eggs spotted and marked with sepia, chiefly around the larger end. These birds of many names are not found in Germany, any more than the so called English snipe is found in England, but they are called German snipe or Deutschers, to distinguish them from that species, dowitcher being simply a corruption of Deutscher in the mouths of longshoremen.

. . . .

The **Long-billed** or **Western Dowitcher** *(Macrorhamphus scolopaceus)*, the representative of the preceding species from the Mississippi Valley westward to Alaska, may be distinguished from it chiefly by its slightly larger size and longer bill and possibly by its more uniformly rusty under parts and the heavier dusky bars on its sides in the summer plumage only. Very rarely one of these birds is taken by gunners on the Atlantic coast. In habits these two species are similar—even their eggs being identical; but the shrill whistled *p'te-te-te, p'te-te-te*, of the gray snipe swells into a musical song, something like *peet-peet; pee-ter-wee-too; wee-too;* twice repeated, according to Mr. D. G. Elliot, in the case of the long-billed dowitcher. For years even scientific men thought these two species were one.

Stilt Sandpiper

(Micropalama himantopus)

Called also: LONG-LEGGED SANDPIPER.

Length—About 9 inches.

Male and Female: *In Summer*—Feathers on upper parts blackish, each bordered with gray or buff or tawny, the markings scalloped on the shoulders; wings darker; ears, and an indistinct line around back of head, rusty red; lower back ashy; upper tail coverts white with dusky bars; tail ashy, the centre and edges of the feathers white. Under parts white, streaked and barred with dusky. Bill nearly as long as a snipe's, and flattened and pitted at the tip. Legs very long. Both bill and feet greenish black. *In Winter:* Upper parts brownish or ashy gray, the feathers edged with white; a white line like an eyebrow, upper tail coverts white, the tail feathers white margined with brownish ash; throat and sides streaked with gray; under parts white.

Range—Eastern North America, nesting within the Arctic Circle, wintering from Florida and the Gulf States to Brazil and Peru.
Season—Spring and autumn migrant, May; July to October.

From the Arctic Circle to Peru is surely a journey to warrant frequent and long breaks; but only rarely do we hear of a small, open flock of these tireless travellers resting awhile on the sand flats of our coast or the muddy channels of the rivers inland to fortify themselves with a square meal before continuing their rapid flight. Like most birds that spend part of their lives at least in Arctic desolation, these sandpipers, not knowing man, have little fear of him, being of the same gentle, confiding disposition, apparently, as the dowitchers, with which they may sometimes be found, lured by the sportsman's decoys. Four birds, watched on a Long Island beach, were wading about in a pool left by the receding tide; and as they tipped forward, thrusting their sensitive bills into the soft sand to feel after food, and often immersing their heads to secure a worm or snail buried there, it seemed as if the top-heavy little waders must upset from their long, slender props. Yet when they walked—for they do not run as actively as true sandpipers, this species being a connecting link between sandpipers and snipe—they moved gracefully and easily. One characteristic they have that reminds one of the avocet and black-necked stilt: on alighting they first teeter, then stand motionless as if to steady themselves and make sure of their balance. Colonel Goss tells of their squatting to avoid detection, flying only as a last resort, then darting swiftly away, calling a sharp *tweet, tweet.*

Knot

(Tringa canutus)

Called also: ROBIN SNIPE OR SANDPIPER; RED-BREASTED SANDPIPER (summer); GRAY SNIPE (winter); BEACH ROBIN; ASH-COLORED SANDPIPER; GRAY BACK.

Length—10.50 inches: largest of the sandpipers.
Male and Female: In summer—Upper parts varied black, gray and reddish; crown gray streaked with black; line over the eye, chin, throat, and underneath cinnamon red, fading to white

on centre of abdomen; rump, upper and under tail coverts and flanks white, barred with dusky; tail ashy brown, bounded by dusky brown and tipped with white. Bill, legs, and feet black. *In winter*—Top of head and back of neck brown, streaked with soiled white; back and shoulders ashy gray, the feathers edged with a lighter shade or white; under parts white, the neck and breast spotted and barred with gray.

Range—Nearly cosmopolitan; nesting in the northern half of the globe and migrating to the southern half in winter. In the United States more common, during the migrations, along the sea coasts than in the Mississippi valley route southward.

Season—Spring and autumn migrant; May and June; July to November.

Like King Canute, this beach robin that Linnæus named for him seems to defy the waves, as, running out after them, it would fain bid them keep back until it has had its fill of the small shellfish left uncovered on the sand; but more quickly running in again when the surf combs and breaks in a threatening deluge. Now it runs nimbly out in the wake of the receding waters, apparently intent only on its dinner, but all the while watching out of the corner of its eye an incoming wave, whose march and volume it so accurately estimates. It is amazing how closely and yet how certainly it escapes a drenching: the tumbling surf never quite overtakes it on its race back, though that last morsel it stopped for seemed inevitably fatal. It is a fascinating, though a nervous, sort of occupation, watching the sandpipers picking up their hurriedly interrupted meals. Drayton gives a different reason for fastening Canute's name on the knot, than the one popularly supposed to be the right one, in his lines:

> "The Knot that calléd was Canute's bird of old,
> Of that great king of Danes his name that still doth hold,
> *His appetite to please*, that far and near was sought."

Not all the knot's food is picked off the surface: the worm, snail, or small crustacean that has buried itself in the soft mud must be probed for, snipe fashion.

Gentle, easily decoyed birds, owing to their fondness for society, usually a good sized bunch, if any, settles down on the mud flat or sandy beach after a preliminary wheel in close array; hence the all too frequent possibility of a single discharge killing the entire company. The marvel is that there are any knots left to shoot. Mr. George H. Mackay, in *The Auk*, tells of the "fire lighting" method of capturing them, once in vogue, which was "for two men to start out after dark at half tide, one of them to carry a lighted lantern, the other to reach and seize the birds, bite their necks, and put them in a bag slung over the shoulder." Sportsmen put a stop to the burning of marshes some years ago, but not until this fine game bird, with many others, had become rare. The same authority quoted describes its notes as "a soft *wah-quoit*, and a little *bonk*." In Kansas, Ohio, and other parts of the interior, where there is no surf to chase out and run from, one meets scattered flocks pattering about on the muddy shores of lakes and rivers, quite as actively as if the water pursued them. Alighting one minute, flying off the next, resting an instant, then on again after a quick little run, the knot sometimes acts more like a fugitive from justice than an inoffensive, peaceful lover of its kind. This restlessness is not so noticeable in the autumn migration, perhaps, when the birds are fat from abundant food, as in the spring, when they make short pauses on the long trip, impatient to reach their nesting grounds within the Arctic Circle.

It was General Greely who first made known the eggs and nest of these birds. "They arrived on June 3, 1883," he writes in his "Three Years of Arctic Service," "and immediately nested (near Fort Conger). . . . The ground color (of the egg) was light pea-green, closely spotted with brown in small specks about the size of the head of an ordinary pin. . . . Fielden has described the soaring of these birds, and the peculiar whirring noise they make."

. . . .

The Purple Sandpiper, Winter or Rock Snipe *(Tringa maritima)*, an extremely northern species, also observed by General Greely near Thank God Harbor, comes down our Atlantic coast between November and March, but not often farther than Long Island or the Great Lakes. Like the Pilgrim Fathers, it chooses to dwell on a "stern and rock bound coast." It is wonderfully sure-footed in running over the slippery bowlders dashed by the spray, picking its food as it goes from among the algæ attached to the rocks. It is nine inches long,

Snipe, Sandpipers, etc.

and in its winter plumage—the only dress we see—the purplish gloss on the black feathers of its back, worn in summer, is not visible. Instead, it is a uniform lustrous ash on its head, neck, breast, and sides. The back, which is a dingy olive brown, has the feathers margined with ash. The wings are the same shade, but the coverts and some of the long feathers are distinctly bordered with white; linings of the wings and under parts are white; the upper tail coverts and middle tail feathers are blackish; the outer feathers, ashy.

Pectoral Sandpiper

(Tringa maculata)

Called also: KRIEKER; JACK, GRASS, COW, and MEADOW "SNIPE"; HAY BIRD; BROWN BIRD; SHORT NECK.

Length—9.00 to 9.50 inches.

Male and Female—The blackish brown feathers of upper parts heavily bordered with buff; the lower back and upper tail coverts black, lightly tipped with buff. Tail pointed; the shorter outer feathers brownish gray, edged with white. Eyebrow white; sides of head, neck, and breast white, streaked with brown or black; rest of under parts white. In winter plumage the feathers of upper parts are edged with chestnut, instead of buff, and the breast is washed with yellow.

Range—The whole of North and the greater part of South America; also the West Indies. Nests in the Arctic regions; winters south of United States.

Season—Migratory visitor, April, May, and from July to November.

To all except inveterate gunners the habits of this little game bird become most interesting after it has gone to the far north, where most people may not observe them, and we must depend upon Mr. Nelson's "Report on Natural History Collections made in Alaska" for our information. On reaching the nesting grounds a male becomes intensely excited in its efforts to win the attention of a sweetheart. It may "frequently be seen running along the ground, close to the female," he writes, "its enormous sac inflated, and its head drawn back and the bill pointing directly forward; or, filled with springtime vigor, the bird flits, with

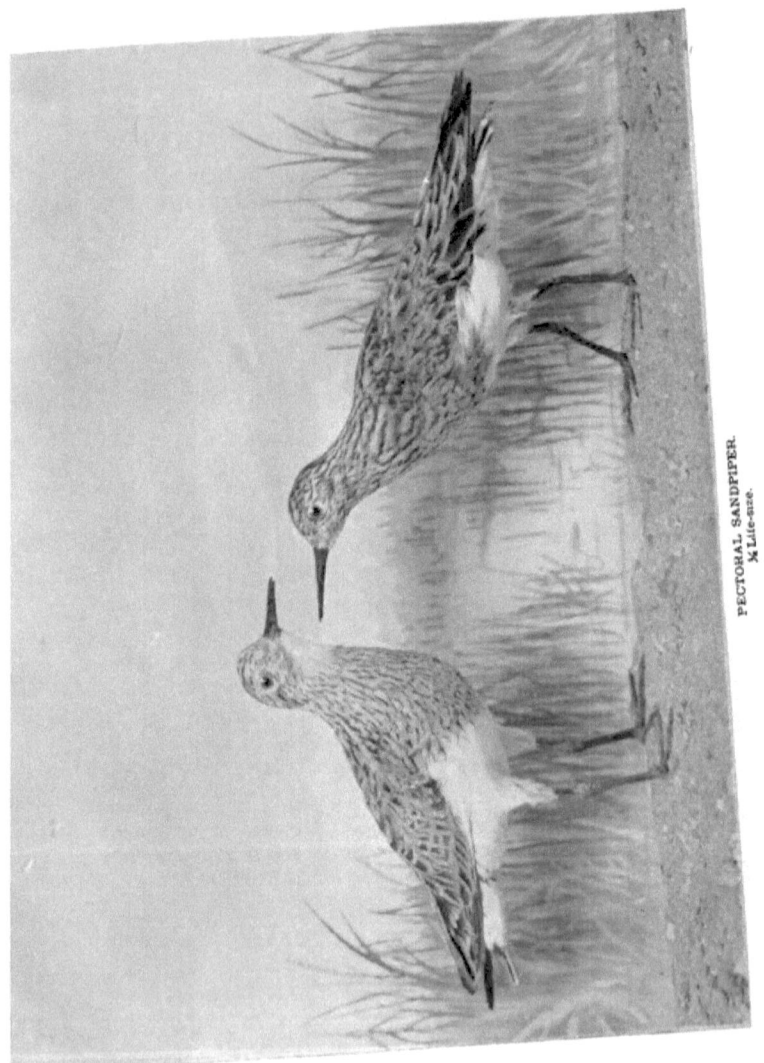

PECTORAL SANDPIPER.
½ Life-size.

slow but energetic wing-strokes, close to the ground, its head raised high over the shoulders, and the tail hanging almost directly down. As it thus flies, it utters a succession of hollow, booming notes, which have a strange ventriloquial quality. At times the male rises twenty or thirty yards in the air, and, inflating its throat, glides down to the ground with its sac hanging below. Again he crosses back and forth in front of the female, puffing his breast out, bowing from side to side, running here and there. . . . Whenever he pursues his love making, his rather low but pervading note swells and dies in musical cadences." These liquid notes may be represented by a repetition of the syllables *too-u, too-u, too-u*. Like certain members of the grouse family, the skin of the throat and breast of the male becomes very loose and flabby, like a dewlap, during the mating season, and may be inflated at will to a size equalling that of the body. Eggs brought to the Smithsonian Institution from tufts of grass in meadows at the delta of the Yukon are greenish drab, spotted and blotched with umber.

When flocks of these sandpipers come down from Alaska and Greenland in early autumn, we see them less commonly scattered on the beaches, where one naturally looks for sandpipers, and usually in the salt marshes, or in meadows near water, salt or fresh, running nimbly among the grasses, pattering about in the pools, pecking at insects, snails, and other tiny creatures above ground, or probing the soft mud or sand for such as have buried themselves below. Silent, gentle, almost tame, friendly with their allies and unsuspicious of foes, they lie well to a dog, squat when danger comes near, and only when it positively threatens fly off with a "squeaky, grating whistle." Because they fly in a zig-zag, erratic course, they are frequently called snipe, but they are true sandpipers, nevertheless. Decoys rarely lure them, though an imitation of their whistle may. In autumn we can see no indication of the extraordinary pectoral sac that becomes so prominent in the bird's figure in June, and that is responsible for the most characteristic of its many popular names.

. . . .

The White-rumped, Schinz's, or Bonaparte's Sandpiper (*Tringa fuscicollis*), scarcely over seven inches long, looks like a smaller copy of the preceding species, although on close scrutiny we note that its central tail feathers are not long and sharply pointed, and that its longer upper tail coverts are white instead of blackish. These white tail coverts, so conspicuous in flight, help to define the bird from Baird's Sandpiper, that has dingy olive brown coverts; but we must depend upon the white-rumped bird's larger size, chiefly, to tell it from the semipalmated sandpiper. This is a sociable little wader, often flocking with its cousins, and so offering frequent opportunities for comparison of these often confused species. In winter the upper parts are plain brownish gray, and the streaks on neck, breast, and sides are less distinctly streaked. No striking peculiarities of habit distinguish it: it is a peaceful, gentle, friendly, active, little sprite, like the majority of its kin; too confiding, often, to save its body from the ultimate fate of the gridiron and the skewer. Its note is a piped *weet, weet.*

. . . .

Baird's Sandpiper (*Tringa bairdii*), far more common in the interior than on the Atlantic coast, closely resembles the white-rumped species in size and plumage, and may be distinguished from it "by the fuscous instead of white middle upper tail-coverts," says Mr. Frank Chapman. "In summer it differs also in the absence of rufous above, the less heavily spotted throat, and the white instead of spotted sides. In winter the chief distinguishing marks of the two species, aside from the differently colored upper tail-coverts, are the buffy breast and generally paler upper parts of *bairdii*." Colonel Goss says these sandpipers are more inclined to wander from the water's edge than the white-rumped species, whose habits they otherwise closely resemble, and that he has flushed them on high prairie lands at least a mile from the water.

Least Sandpiper
(Tringa minutilla)

Called also: PEEP; MEADOW OX-EYE; STINT; WILSON'S STINT; SANDPEEP

Length—6 inches. Smallest of our sandpipers.

Male and Female—*In summer:* Upper parts dingy brown, the feathers edged with chestnut or buff; the lower back and upper tail coverts plain black, like the central tail feathers; outer tail feathers ashy gray. Line over eye, throat, sides, and underneath white, more buffy, and distinctly streaked with blackish brown on neck and breast. (Immature birds have not these distinct streaks.) *In winter:* General appearance gray and white; upper parts brownish gray; breast white or pale gray; not distinctly streaked; other parts white. Bill black; legs greenish; toes without webs.

Range—North America at large; nesting in the Arctic regions and wintering from the Gulf states to South America.

Season—Transient visitor; May; July to October.

Flocks of these mites of sandpipers, often travelling with their semipalmated cousins, whose popular names are indiscriminately applied to them also, come out of the far north just as early as the young are able to make the long journey. Chicks that in June leave the drab or yellowish eggs thickly spotted with chestnut brown, run from the mossy ground-nest at once; and in July, when family parties begin to congregate in Labrador, join the whirling companies of adults in many a preliminary wing drill before descending to the States. Innocent of evil, confiding, sociable, lively little peepers, their tiny bodies offering less than a bite to a hungry man, neither their faith in us nor their pathetic smallness protects them from the pot hunters. True sportsmen scorn to touch them. A single pot shot may and usually does kill a score of birds; yet, so ignorant are they of man and his inventions, the startling report of a gun drives them upward but a few yards for a confused whirl *en masse* that ends on the ground where it began, and often before the dead and wounded victims can be picked up. Celia Thaxter's lines on the little sandpiper charmingly describe its touching confidence.

Snipe, Sandpipers, etc.

Running nimbly along the mud and sand flats of beaches; over rocks slippery with seaweed; in marshes and dry, grassy inland meadows too; or dancing just in advance of the frothing ripples, where the waves break high on the sand, graceful and dainty in every movement they make, these tiny beach birds enliven our waste places until November storms drive them south. Who cannot recall a walk along some beach made memorable by the cheerful companionship of these gay mites running and flitting not far ahead and calling back *peep, peep,* in response to one's whistle? By far the most numerous waders that visit us, one can scarcely fail to find them, if not in scattered companies apart, then in flocks of their numerous relations. Usually they are busily, playfully gathering larvæ, insects, worms, and tiny shell fish that may be picked off the surface or probed for, a quiet intruder not in the least interrupting their dinner. Startle them and they gather into a mass, whirling about, showing their backs as well as their under parts, and with much shrill *peeping;* but their easily restored confidence soon returns, and they again alight on the good feeding ground, though it may not be a rod away.

. . . .

The Semipalmated (half webbed) Sandpiper, or Sand Ox-eye, also known as Peep (*Ereunetes pusillus*), scarcely more than a half inch longer than the least sandpiper, and so like it in plumage and habit it may scarcely be distinguished from it in a flock where these two cousins mingle, has its toes half webbed, its diagnostic feature. Those who refuse to shoot birds in order to name them will have some difficulty here. Possibly this sandpiper keeps closer to the water than its little double that is often found in the meadows. Both birds are so frequently seen chasing out after the waves, to pick up the tiny shell fish, worms, etc., they uncover, and more rapidly being chased in by them as the foam curls around their slender legs, that it is impossible to think of either as anything but beach birds. They are marvelously expert in estimating the second they must run from under the combing wave about to break over their tiny heads; but if the rushing waters threaten a deluge, up they fly, flitting just above the foaming ripples until they subside, leaving a harvest behind. The semipalmated sandpiper swims well when lifted off its feet by an unexpected breaker, or when wounded in the wing.

. . . .

LEAST SANDPIPER.

The Western Semipalmated Sandpiper (*Ereunetes occidentalis*), the representative of the preceding species west of the Mississippi, differs from it in having the plumage of its upper parts more distinctly chestnut red, the breast more heavily streaked, and the bill a trifle longer; but neither species differs perceptibly in habits from the least sandpiper, and neither one is larger than an English sparrow.

Red-backed Sandpiper

(Tringa alpina pacifica)

Called also : DUNLIN; BLACK-BELLIED SANDPIPER; BLACK BREAST; PURRE; FALL OR WINTER SNIPE; LEAD-BACK; BRANT SNIPE; STIB; OX-BIRD

Length—8 to 9 inches.

Male and Female—In summer: Chestnut red streaked with black above, many feathers tipped with white; lower back and upper tail coverts blackish; wing coverts and tail feathers brownish gray; breast whitish streaked with dusky; under parts white, with a large black patch in the middle. (Summer dress worn early and late.) *In winter:* Upper parts brownish or ashy gray; under parts white or grayish, sparingly streaked; the sides sometimes spotted with black. Bill long, black, and curved downward; legs and feet black. Immature birds have the blackish feathers of upper parts with rounded tips of chestnut or buff; the breast washed with buff and indistinctly streaked; white underneath, spotted with black.

Range—North America; nesting in the Arctic regions, wintering from Florida southward. A few remain farther north in sheltered marshes. Rare inland; common coastwise.

Season—Transient visitor; April, May; August to October.

Never far from the sand bars and mud flats exposed at low tide, or the salt water marshes back of the beaches, flocks of these red-backed sandpipers, that are not always clad in their winter feathers when they come to spend the autumn on our shores, pursue the daily round of duties and pleasures common to their tribe. It is not an easy matter, even to one well up in field practice, to name the multitudinous sandpipers on sight, since their plumage, never bold or striking, often differs greatly with age and season, making the task even more difficult than that of correctly naming every warbler. But the long, decurved bill of this sandpiper offers the surest clue to its identity at any time.

With this bill the sand worms are dragged forth from their holes and the tiny shell fish from the depths in which they have buried themselves at low tide. It appears to be quite as sensitive in feeling after food as a snipe's. Or it will be used to pick morsels from the surface and to seize insects on the wing in the salt meadows. Usually these sandpipers keep close together in their feeding grounds and during flight, offering all too tempting a chance for a pot shot. Because they are unsuspicious from passing so much of their lives in Arctic desolation, unmolested by men, dogs, and guns, their gentle confidence passes for stupidity here. Is it through stupidity or some higher trait that the survivors of a flock, just raked by a bayman, return immediately, after a hurried, startled whirl, to the spot where their companions lie dead or wounded and helpless, calling forth a pity in them not shared by the man behind the gun, who, with another discharge, rakes the survivors? One inveterate old reprobate on Long Island proudly exhibited over fifty of these and pectoral sandpipers that had been feeding with them, as victims of only three shots.

In the spring, when lively impulses move all birds to interesting performances, these dunlins, as our English cousins call them, go through some beautiful wing manœuvres calculated to inspire admiration in the speckled breast of the well beloved. "As the lover's suit approaches its end," to cite an author quoted by Mr. D. G. Elliot, "the handsome suitor becomes exalted, and in his moments of excitement he rises fifteen or twenty yards, and hovering on tremulous wings over the object of his passion, pours forth a perfect gush of music until he glides back to earth exhausted, but ready to repeat the effort a few minutes later. Murdoch says their rolling call is heard all over the tundra every day in June, and reminds one of the notes of the frogs in New England in spring." Up at the far north, where the love making and nesting are commenced by the first day of summer—for the birds make a very short stay here in spring—the males utter "a musical trilling note, which falls upon the ear like the mellow trickle of large drops falling rapidly into a partly filled vessel." Three or four precocious chicks, that have emerged from pale bluish white or buff shells heavily marked with chocolate, run about the tundra with their still devoted parents in June, and are able to fly expertly in July, when the first migrants reach our shores.

Sanderling

(Calidris arenaria)

Called also: SURF SNIPE; RUDDY PLOVER; BEACH BIRD

Length—7 to 8 inches.

Male and Female—*In summer:* Upper parts varied blackish brown, reddish chestnut, and grayish white, most feathers tipped with the latter; wing coverts ashy brown, broadly tipped with white, making a bar across wings; tail brownish gray, margined with white, the outer feathers nearly white; throat and breast washed with pale cinnamon and spotted with blackish, other under parts, immaculate white. Bill, about as long as head, stout, straight, black; broader at the tip than at its slightly concave centre. Feet with three toes only; no hind toe; scales of tarsus transverse. *In winter* The chestnut in upper plumage replaced by gray, or mixed with brown and gray in the spring; under parts pure white. Immature birds in autumn lack the chestnut tint and are more evenly mottled; brownish ash or blackish and white above, pure white below; rarely with a spot on breast.

Range—Nearly cosmopolitan, nesting in the Arctic regions or near them; south in winter as far as Chile and Patagonia.

Season—Spring and autumn visitor; March to June; September, October.

Commonest of the beach birds everywhere, the sanderlings —for it is impossible to think of them except in flocks—run about like a company of busy ants on our coast and sometimes inland too, near large bodies of water that are followed in the migrations. Gleaning from the sand flats with an eagerness suggesting starvation, their heads pushed forward, alert, nimble-footed, nervously quick in every movement, the birds' every energy while with us appears to be concentrated on the business of picking up a living as if they never expected to see food again.

Snipe, Sandpipers, etc.

Among the semipalmated, the least, and other sandpipers they often hunt with, sanderlings may be readily picked out by the attitude of the head and their fearful eagerness. Impressions of their three toes (a plover characteristic) in the wet sand, at low tide, cover a good feeding ground like fret work. Chasing out after the receding breakers, picking up the minute shell fish, marine insects, shrimps, seeds of sedges, etc., strewn over the flats, the active little troop outstrips the frothing waves on the backward race with marvelous agility. Rarely, indeed, does the curling foam reach the immaculate white under plumage; no combing breaker ever drenches the sanderlings unawares, however absorbing their dinner appears to be; yet deep water has no terrors for them. Wading is a frequent diversion, and swimming becomes the safest resort for wounded birds.

Bay men, who habitually carry guns and shoot at everything wearing feathers, tell you that sanderlings are wary little creatures, never so gentle and confiding as many sandpipers that may be raked from a few yards; but possibly if these men carried only field glasses, and kept up a reassuring *peet-meet* whistle as they slowly approached a busy flock—a possibility to make a longshoreman smile—the alleged timidity would be found to disappear and the birds to remain. Startle them, and rising and moving like one bird toward the sea, calling shrilly as they fly, on they go along the coast line no further than a few hundred yards, their bodies turning and twisting in the air, their under parts glistening where the sunlight strikes them. Instantly, on alighting, the flock begins to feed again. Follow these birds to Florida in winter, and one finds apparently the same ones still feeding. Captain Feilden, the naturalist in General Greely's Arctic expedition, reported sanderlings in flocks of knots and turnstones, and a nest in latitude 82° 33' north. It was on a gravel ridge above the sea, and the eggs (three or four light olive brown, finely spotted and speckled with darker) were deposited in a slight depression among ground willow plants, the lining of the nest consisting of a few withered leaves and dry catkins.

Marbled Godwit

(Limosa fedoa)

Called also: MARLIN; BROWN MARLIN; STRAIGHT-BILLED CURLEW; RED CURLEW; GREAT MARBLED GODWIT; DOE-BIRD

Length—16 to 22 inches; largest of the shore birds except the long-billed curlew.

Male and Female—General impression of plumage pale, dull chestnut red barred and varied with black. Head and neck pale buff streaked with black; entire upper parts reddish buff, irregularly barred with black or dusky; throat white; rest of under parts pale reddish buff, the strongest shade under wings; wavy dark brown lines on all feathers except on centre of abdomen, which is pale buff. Long bill, curving slightly upward, flesh colored at base, blackening near the tip; long legs, ashy black. Female larger. Immature birds are similar, but lack most of the brown lines on under parts.

Range—Temperate North America; nesting in the interior chiefly from the upper Mississippi region north to the Saskatchewan; wintering in Cuba, Central and South America; rare on Atlantic coast.

Season—Chiefly a spring and autumn migrant in United States; May; August to November.

Conspicuous by its size and coloration among the waders, the great marbled godwit might be confused only with the long-billed curlew at a distance where the slight curve upward of the godwit's bill and the pronounced downward curve of the curlew's could not be noted. It is not the intention of the godwit to give anyone a near view of either plumage or bill. The most stealthy intruder on its domains—salt or fresh water shores, marshes, and prairie lands—startles it to wing; its loud, whistled notes sound the alarm to other marlins hidden among the tall sedges, and the entire flock flies off at an easy, steady pace, not rapid, yet not to be overtaken afoot. A beautiful posture, common to the plovers, curlews, terns, and some other birds, is struck just as they alight. Raising the tips of the wings till they meet high above the back, the marlins suggest the favorite attitude of angels shown by the early Italian painters.

Devoted to their companions, as most birds of this order are,

Snipe, Sandpipers, etc.

the godwits lose all shyness and caution when some members of the flock that have been wounded by the gunner, cry out for help. Unwilling to leave the place, and hovering round and round the spot where a dead or dying comrade lies, they seem to forget their fear of men and guns, now replaced by a sympathy that risks life itself. Just so they hover about a nest and cry out sharply in the greatest distress when it is approached, until one feels ashamed to torment them by taking a peep at the four clay colored eggs, spotted, blotched, and scrawled over with grayish brown, where they lie in a grass lined depression of the ground. The nests are by no means always near water; several seen in Minnesota were in dry prairie land.

The marlin feels along the shore somewhat as an avocet does, its sensitive bill thrust forward almost at a horizontal, as touch aids sight in the search for worms, snails, small crustaceans, larvæ, and such food as may be picked off the surface or probed for as the bird walks along. Suddenly it will stop, thrust its bill into the mud or sand up to the nostrils, and, snipe fashion, feel about for a worm that has buried itself, but not escaped. Standing on one long leg, the other somehow concealed under the plumage, the neck so drawn in it seems to be missing from the marlin's anatomy, the bill held at a horizontal—this is a characteristic attitude whether the bird be standing knee deep in the water, or among the prairie grass.

The **Hudsonian Godwit, Ring-tailed Marlin, White-rumped, Rose-breasted,** or **Red-breasted** Godwit (*Limosa hæmastica*), while it resembles the preceding in habits, differs from it in length, which is about fifteen inches, and in plumage, which is as follows: Upper parts black or dusky; the head and neck streaked with buff, the back barred or mottled with it; upper tail coverts white (conspicuous in flight), the lateral coverts tipped or barred with black; the tail black, with a broad white base and narrow white tip; throat buff streaked with dusky; the under parts chestnut red barred with black, and sometimes tipped with white. This bird, not so rare on the Atlantic coast as its relative, is nevertheless not common either there or elsewhere in the United States.

Greater Yellowlegs

(Totanus melanoleucus)

Called also: BIG YELLOWLEG; TELLTALE OR TELLTALE SNIPE; LARGE CU-CU; YELPER; TATTLER; STONE SNIPE; WINTER YELLOWLEG; YELLOW SHANKS

Length—13 to 14 inches.

Male and Female—Upper parts dark ashy speckled with white; the head and neck streaked; the back and wings spotted; space over eye and the throat white; tail dusky, with numerous white bars; the white breast heavily spotted with black; sides barred; underneath plain white. (Winter and immature birds have the upper parts more ashy or gray, and almost black in summer, and the markings on sides and breast fade in autumn.) Bill two inches long or over; long, slender, yellow legs.

Range—America in general, nesting from Iowa and northern Illinois northward, and wintering from the Gulf states to Patagonia.

Season—Chiefly a spring and autumn visitor; April, May; July to November.

A "flute-like whistle, *wheu, wheu-wheu-wheu-wheu, wheu, wheu-wheu*," familiar music to the sportsmen in the marshes, tells the tale of the yellowlegs' whereabouts; and a responsive whistle, calling down the noisy, sociable birds to the wooden decoys even from a greater height than their bodies may at first be seen, or bringing them running from the muddy feeding grounds to their supposed friends, lures them close enough to the blind for a pot shot. Consternation seizes the survivors; they fly upward and jostle against each other; they dart now this way, now that, crying shrilly as they blunder upward in a zigzag course; but calming their fears as the whistle from behind the blind reassures and entreats, down wheel the confiding innocents again, only to be stretched beside their stiffening companions at a second discharge of the gun. So this alleged sport goes gaily on through the autumn, although no one on the Atlantic coast, at least, raves over the sedgy flavor of the stone snipe's flesh, or often tries to give a better reason for bagging the birds than that they frighten off the ducks! In the west the flesh is more truly desirable.

Snipe, Sandpipers, etc.

Noisy, hilarious chatterers, their shrill notes, **four times repeated,** coming from an entire flock at once, after **the** manner of old squaws, these tattlers, that are always inviting kindred flocks to join theirs, excite other birds to restless habits like their own, and keep themselves well advertised in the marshes and about the bays and estuaries where they feed. Yet they are exceedingly vigilant in spite of their noise, and are the first to pass an alarm. It is only **by screening oneself behind a blind,** and whistling the birds within range of nothing more **formidable** than a field glass and **a camera,** that the altruistic bird **hunter may** hope to study the wary fellows. As a flock whirls about **in wide, easy** circles before alighting, they appear to be yellow **legged white birds.** Before actually touching the ground with their dangling **feet, the** wings are flapped, then raised above the back to a point where they meet—a posture suggesting a scorn of earth—then they are softly folded into place. As the bird walks, it carries itself with a stately dignity, yet the long bill turned inquisitively from side to side detracts not a little from the general impression of elegance. Wading up to its breast in shallow waters, or running nimbly over the sand flats and muddy beaches, the yellowleg keeps its bill almost constantly employed **dragging** worms, snails, and small shell fish from **their holes, probing for others,** and picking **up tiny** crustaceans **swimming along the** surface of the water or crawling over the beach.

It is a long excursion from Labrador to the Argentine Republic, yet birds hatched at the end of June at the north reach South America in October, leaving again in March, and so enjoy perpetual summer.

Yellowlegs

(Totanus flavipes)

Called also: SUMMER YELLOWLEGS; LESSER TELLTALE; TELLTALE SNIPE; YELPER; LITTLE CU-CU; LESSER YELLOWSHINS; LITTLE STONE SNIPE, ETC

Length—10 to 12 inches.

Male and Female—Coloration precisely as in the greater yellowlegs. This bird is to be distinguished only by its smaller size, and its proportionately longer legs.

YELLOW LEGS.
½ Life size.

Range—North America at large, nesting from the northern states to the Arctic regions; wintering from the Gulf states to Patagonia.
Season—Chiefly a spring and autumn visitor; more abundant in autumn; rarely a summer resident; April, May; July to October.

The haunts, habits, and noisy voices of the two species of yellowlegs are so nearly identical, like their plumage, that a description of them would be simply a repetition of the larger bird's biography. From the fact that some of these birds nest within the United States limits, they have been called summer yellowlegs; but the great majority act precisely as their larger double does, and so have earned only diminutives of its popular names. In the Mississippi region the lesser telltale is far more common than in the east, but it is still abundant on the Atlantic coast in the autumn migrations, at least; and it is supposed to be everywhere a commoner bird than the greater yellowlegs. Possibly this smaller tattler responds more readily to the whistling down method of enticing a flock to decoys, but the experiences of individual sportsmen differ greatly in this as in most matters.

Solitary Sandpiper

(Totanus solitarius)

Called also: GREEN SANDPIPER; SOLITARY, or WOOD TATTLER

Length—8 to 9 inches.
Male and Female—*In summer:* Upper parts dingy olive with a greenish tinge, streaked on the head and neck, and finely spotted on the back with white; tail regularly barred with black and white, the white prevailing on the outer feathers; primaries and edge of wing blackish; underneath white, shaded with dusky and streaked on sides of throat and breast; sides and wing linings regularly barred with dusky. Long, slender, dark bill; legs dull green, turning black after death. *In winter:* Similar, but upper parts more grayish brown and the markings everywhere less distinct.
Range—North America, nesting occasionally in northern United States, but more commonly northward, and migrating southward in winter to Argentine Republic and Peru.

Snipe, Sandpipers, etc.

Season—Spring and autumn visitor; April, May; July to November. Rarely a summer resident.

A lover of woods, wet meadows, and secluded inland ponds, in the lowlands or the mountains rather than the salt water marshes and sand flats of the coast that most of its kin delight in, the wood tattler is a shy recluse, but not a hermit. At least a pair of birds are usually seen together, representatives of small flocks scattered over the neighborhood, but generally hidden in the underbrush. As compared with most other sandpipers that move in compact flocks and are ever inviting other waders to join them, this species is certainly unsocial; but to call it solitary implies that it is a misanthrope like the bittern, which no one knew better than Wilson, who named it, that it is not. "It is not a morose or monkish species, shunning its kind," says Mr. D. G. Elliot, "but is frequently met with in small companies of five or six individuals on the banks of some quiet pool in a secluded grove, peacefully gleaning a meal from the yielding soil or surface of the placid water. As they move with a sedate walk about their chosen retreat, each bows gravely to the other, as though expressing a hope that his friend is enjoying most excellent health, or else apologizing for intruding upon so charming a retreat and such select company." Dainty, exquisite, graceful, exceedingly quick in their movements, their chief fault is in keeping out of sight so much of the time—the characteristic that preserves their delicate flesh from overloading game bags. Penetrate to their retreats, and they prefer running into the underbrush rather than expose their neat figures and speckled plumage by skimming over the pond. Sit down on the bank, and perhaps some dapper little fellow will pay no attention to your motionless figure and pursue his own concerns. He will run nimbly along the margin of the water, snapping at insects and caterpillars here and there, or, rising lightly in the air, seize a small dragonfly on the wing. He may go lightly over the lily-pads, rail fashion, half flitting with his wings, half running to keep himself from sinking, or wade up to his breast with measured steps, heron fashion, and remain fixed there, waiting for the small coleoptera to skip along the surface within range of his bill. This species appears to eat comparatively few snails, worms, and crustaceans, and a preponderance of insect fare. Its low, musical whistle is rarely heard here, but the South Americans see the propriety of calling this bird a tattler.

Although the solitary sandpiper is known to make its nest in the United States, so cleverly does it conceal it, only a single clutch of eggs has ever been found, so far as known, the one taken by Richardson near Lake Bombazine, Vermont, in May, 1878. Dr. Brewer described the eggs as light drab, with small rounded brown markings, some quite dark, nowhere confluent, and at the larger end a few faint purplish shell marks.

Willet

(Symphemia semipalmata)

Called also: SEMIPALMATED TATTLER

Length—About 16 inches.

Male and Female—*In summer:* Upper parts brownish gray, streaked on the head and neck with black; the back barred across with black, which sometimes give the prevailing tone; a large white space on wings, half the primaries and the greater part of secondaries being white; upper tail coverts white, indistinctly barred with dusky; central tail feathers ashy, indistinctly barred with dusky; the outer feathers almost white, and mottled with gray. Under parts white; the fore neck heavily streaked; the breast and sides washed with buff and heavily barred with dusky; wing lining sooty. Bill long and dark; legs bluish gray; the toes partly webbed (semipalmate). *In winter:* Upper parts a lighter brownish gray, nearly if not altogether unmarked; the tail coverts white; below white shaded with gray on throat, breast, and sides; axillars blackish. A great variety of intermediate stages.

Range—Eastern temperate North America, nesting throughout its United States range, but rarely north of Long Island or Illinois; resident in southern states, and wintering southward to West Indies and Brazil.

Season—Summer resident or spring and autumn visitor; May; August and September.

Pill-will-willet, pill-will-willet, loudly whistled from the tide or fresh water marshes, leaves no doubt in the sportsman's mind as to what bird is sounding the alarm to better game and startling every throat and wing in the neighborhood to action. Wary, restless, noisy, no one may approach this large tattler, however well protected in the spring, at least (as every bird should be), under the wing of the law; neither will it come to a decoy easily, nor permit itself to be whistled down to the stools, unlike the majority of its too confiding kin. But however distrustful of man, it is not unsocial, since we often see it in companies of other beach birds that evidently depend upon its office as sentinel. Morning, noon, and night its voice is loudly in evidence, until one tires of hearing its persistent whistle. Within a stone's throw of a summer cottage on the New Jersey coast, a decidedly wide-awake call came from the marsh every hour between sunset and sunrise.

But love, the magician, works wonders with this noisy, distrustful bird, and a radical if temporary change comes over it during the nesting season. "They cease their cries," says Dr. Coues, "grow less uneasy, become gentle, if still suspicious, and may generally be seen stalking quietly about the nest. When willets are found in that humor—absent minded, as it were, absorbed in reflection upon their engrossing duties, and unlikely to observe anything not in front of their bill—it is pretty good evidence that they have a nest hard by. During incubation, the bird that is 'off duty' (both birds are said to take turns at this) almost always indulges in reveries, doubtless rose tinted . . . and the inquiring ornithologist could desire no better opportunity to observe every motion and attitude."

A nest in the Jersey marsh already mentioned was nothing more than a depression in a dry spot of ground, containing four pale olive brown eggs spotted with a darker shade and rich purplish brown. This nest, among the thick sedges, was reached by a sort of tunnel among the grasses, entered some little distance away by the sitting bird. Neither parent had forgotten how to get scared or to make a noise the day that nest was visited; nor did other birds in the marsh fail to loudly protest their sympathy, not to say alarm, as they circled overhead in a state of painful excitement. Reassured that no harm had been done by a mere glance at the speckled treasures, the willets wheeled about lower and lower over the sedges, flashing the white wing mirrors in the sunlight before they alighted, and with wings held high above the back until they met, at last set foot to earth again, bowing their heads like reverent archangels as they struck this exquisite posture. Musical, liquid, tender notes, evidently a love song, float from the throat of the sentinel lover, walking up and down in absent-minded happiness not many paces from the entrance to the grassy tunnel. None of the willets in that well populated marsh were ever caught in the act of swimming, though the partial webbing of their feet indicates that they must be able to swim well when necessary. A western representative of these birds, formerly confounded with them, nests west of the Mississippi, and Mr. William Brewster discovered that it is a slightly larger bird, with a more slender, long bill, of paler coloration, and with less distinct bars and other marks.

Bartramian Sandpiper
(Bartramia longicauda)

Called also: UPLAND, or FIELD, or GRASS, or HIGHLAND PLOVER; BARTRAM'S TATTLER; PRAIRIE PIGEON; PRAIRIE SNIPE; QUAILY

Length—11.50 to 12.75 inches; usually just a foot long.

Male and Female—Upper parts blackish varied with buff, brown, and gray; the head and neck black streaked with buff, and a buff stripe through the eye; the back and the wing coverts dusky barred with buff, the lighter color prevailing on the nape and wings; outer primary olive brown barred with white, the others barred with black; lower back, rump, and central tail coverts brownish black; tail feathers brownish gray, the outer ones varying from orange brown to buff or white, all more or less barred with black, with a broad black band across the end, and white tips of increasing breadth. Under parts white, washed with buff on breast and sides, which are streaked or barred with black. Bill comparatively short, yellow, with black ridge and tip; feet dull yellow.

Range—North America, chiefly east of Rocky mountains and north to Nova Scotia and Alaska; nesting nearly throughout its North American range; wintering southward so far as Brazil and Peru.

Season—Summer resident or migrant; April, July, August, September.

It is in high, dry, grassy meadows, among the stubble in old pastures, in rustling corn fields and on the open plains, and not always near salt water, that the sportsman looks for this so called wader, more precious in his sight than any other small game bird

Snipe, Sandpipers, etc.

except possibly the woodcock, Bob White, and Jack snipe. Few birds have been more tirelessly sought after; few that were ever abundant in New England and other eastern states have been so nearly exterminated there by unchecked, unintelligent, wanton shooting. It is to Kansas, Texas, and the great plains watered by the Missouri that one must now go to find flocks numbering even fifty birds, whereas our grandfathers once saw them in flocks of thousands on the Atlantic slope. Like the geese, ducks, and certain other birds that are exceedingly afraid of men and impossible to stalk afoot, this wary "plover" pays no attention whatever to horses and cattle; hence shooting from a wagon is the common method of hunting it in some parts of the west to-day; and an unsuspicious flock, suddenly startled to wing only when the wheels rumble beside it, soon fairly rains plover. Shot easily penetrates the delicate tender flesh unprotected by a dense armor of feathers such as generally saves a goose under similar circumstances.

Delicious as a broiled plover is, there is no true sportsman who will hesitate to admit that the graceful, slender, beautifully marked, sweet voiced bird is not vastly more enjoyable in life. A loud, clear, mellow, rippling whistle that softly penetrates to surprising distances, like the human voice in a whispering gallery, has an almost ventriloqual quality, and one never knows whether to look toward the clouds or among the stubble at one's feet for the musician. For liquid purity of tone can another bird note match this triplet? At the nesting season, especially, a long, loud, weird cry, like the whistling of the wind, *chr-r-r-r-r-e-e-e-e-c-oo-oo-oo-oo*, as Mr. Langille writes it, may be heard even at night; the mournful sound is usually uttered just after the bird has alighted on the ground, fence, or tree, and at the moment when its wings are lifted, till they meet above its back. Everyone who has ever heard this cry counts it among the most remarkable sounds in all nature. The spirited alarm call, *quip-ip-ip*, *quip*, *ip*, *ip*, rapidly uttered when the bird is flushed in its feeding grounds, and still another sound, a discordant scream quickly repeated, that comes chiefly from disturbed nesting birds, complete the list of this tattler's varied vocal accomplishments.

If this upland plover realized how perfectly the plumage on its back imitates the dried grass, it might safely remain motionless and trust to the faultless mimicry of nature to conceal it.

BARTRAMIAN SANDPIPER.
½ Life-size.

As you look down from your saddle into a dry field, the sharpest eye often fails to see these birds squatting there until something (but not the horse) frightens them and a good sized flock surprises you when it takes wing. Three or four sharply whistled notes ring in your very ears as the plovers mount. The swift flight is well sustained. Mere specks seem to float across the heavens, and were it not for the soft, clear rippling whistle that descends from the clouds, who would imagine that these birds so commonly seen on the ground would penetrate to such a height above it? In the migrations along the coast and inland, serried ranks, flying high, cover immense distances daily. The pampas of the Argentine Republic hold flocks that have gathered on our own great plains, who shall say how soon after the journey was begun?

On alighting, with their wings stretched high above their backs in plover fashion, these true sandpipers remain perfectly still for a minute, turning their slender necks now this way, now that, to reconnoitre, before they gracefully walk or run off to feed, bobbing their heads as if satisfied with the prospect as they go. They must devour grasshoppers by the million—another reason why they should be protected. In the nesting season, at least, the mates keep close together when feeding on berries and insects, that, however largely consumed, fail to fatten their slender bodies now. Anxiety, common to all true lovers and devoted parents, keeps them thin. A few blades of dry grass line the merest depression of the ground in some old field or open prairie that answers as a cradle for the four clay colored eggs spotted over with dark brown and clouded with purplish gray shell marks. Funny, top-heavy, fluffy little chicks tumble clumsily about through the grass in June.

. . . .

The Buff-breasted Sandpiper (*Tryngites subruficollis*) closely allied to the larger upland "plover," like it prefers dry fields and grassy prairie lands, although during the migrations it too is often met with on beaches on the coasts of both oceans. Its upper parts are pale clay buff, the centre of each feather black or dark olive; the inner half of the inner webs of the dusky primaries is speckled with black, a diagnostic feature; the longer inner wing coverts are conspicuously marked and tipped with black edged with white; the feathers of under parts are pale buff edged with white and indistinctly marked. A few of these migrants rest awhile on the south shore of Long Island in the early autumn yearly. "They are an extremely active species when on the wing," writes Dr. Hatch, who studied them in Minnesota, "and essentially ploverine in all respects, seeking sandy, barren prairies where they live upon grasshoppers, crickets, and insects generally, and ants and their eggs especially. I have found them repasting upon minute mollusks on the sandy shores of small and shallow ponds, where they were apparently little more suspicious than the solitary sandpipers are notably. The flight is in rather compact form, dipping and rising alternately, and with a disposition to return again to the neighborhood of their former feeding places." "During the breeding season," says Mr. D. G. Elliot, "they indulge in curious movements, one of which is to walk about with one wing stretched out to its fullest extent and held high in the air. Two will spar like fighting-cocks, then tower for about thirty feet with hanging legs. Sometimes one will stretch himself to his full height, spread his wings forward and puff out his throat, at the same time making a clucking noise, while others stand around and admire him."

Spotted Sandpiper
(Actitis macularia)

Called also: TEETER; TILT-UP; SAND LARK; PEET-WEET; TEETER-TAIL

Length—7.50 inches.

Male and Female—Upper parts an olive ashen color, iridescent, and spotted and streaked with black; line over eye and under parts white, the latter plentifully spotted with round black dots large and small, but larger and closer on the male than on the female, the smallest marks on throat; inner tail feathers like the back, the outer ones with blackish bars; secondaries and their coverts broadly tipped with white; some white feathers at bend of wing; white wing lining with dusky bar; other white feathers concealed in folded wing, but conspicuous in flight. Bill flesh colored or partly yellow, black tipped. Winter birds are duller and browner and without bars on upper parts.

Range—North America to Hudson Bay, nesting throughout its range; winters in southern states and southward to Brazil.

Season—Summer resident; April to September or October.

The familiar little spotted sandpiper of ditches and pools, roadside and woodland streams, river shores, creeks, swamps, and wet meadows—of the sea beaches, too, during the migrations, at least—quite as frequently goes to dry uplands, wooded slopes, and mountains so high as the timber line, as if undecided whether to be a shore or a land bird, a wader or a songster. Charming to the eye and ear alike, what possible attraction can a half dozen of these pathetically small bodies roasted and served on a skewer have to a hungry man when beefsteak is twenty cents a pound? A thrush is larger and scarcely more tuneful, yet numbers of these little sandpipers are shot annually.

Some quaint and ridiculous mannerisms, recorded in a large list of popular names, make this a particularly interesting bird to watch. Alighting after a short, low flight, it first stands still, like a willet, to look about; then making a deep bow to the spectator, you might feel complimented by the obeisance, did not the elevation of the rear extremity turned toward you the next minute imply a withering contempt. Bowing first toward you, then from you, the teeter deliberately sea-saws east, west, north, south. This absurd performance, frequently and ever solemnly indulged in, interrupts many a meal and run along the beach. A sudden jerking up or jetting of the tail as the bird walks, like the solitary sandpiper, gives it a most curious gait, all the more amusing because the bird is so small and evidently so self-satisfied. One rarely sees more than a pair of these sandpipers in a neighborhood which they somehow preëmpt, except at the migrations, when families travel together; but as two broods are generally raised in a summer, these family parties are no mean sized flock. Startle a "teeter snipe," and with a sharp, sweet *peet-weet, weet-weet*, it flies off swiftly on a curve, in a steady, low course, but with none of the erratic zig-zags characteristic of a true snipe's motions, and soon alights not far from where it set out. A fence rail, a tree, or even the roofs of outbuildings on the farm have been chosen as resting places. The peet-weets skim above the waving grain inland, their pendant, pointed wings beating steadily, and follow the same graceful curves that mark their course above the sea.

In the nesting season, which practically extends all through the summer, this is a sand "lark" indeed. Soaring upward, singing as he goes, in that angelic manner of the true lark of England, the male pours out his happiness in low, sweet *peet-weets* trilled rapidly and prolonged into a song;—cheerful, even ecstatic notes, without a trace of the plaintive tone heard at other times. A good deal of music passes back and forth from these birds a-wing. Fluffy little chicks run from the creamy buff shells thickly spotted and speckled with brown, as soon as hatched. The nest, or a depression in the ground, lined with dry grass, that answers every purpose, may be in a meadow or orchard, but rarely far from water that attracts worms, snails, and insects for the little family to feed on. This is the one sandpiper that we may confidently expect to meet throughout the summer.

Long-billed Curlew

(Numenius longirostris)

Called also: SICKLE-BILL; SABRE-BILL; SPANISH CURLEW; BUZZARD CURLEW

Length—24 inches; bill of extreme length, about 6 inches, sometimes 8 inches.

Male and Female—Upper parts buff or pale rufous and black; the head and neck streaked; the back, wings, and tail barred or mottled with cinnamon, buff, and blackish; under parts buff; the breast streaked, and the sides often barred with black. Long, black bill, curved downward like a sickle; long legs and feet, dark.

Range—Temperate North America; nesting in the south Atlantic states and in the interior so far as Hudson Bay, or mostly throughout its range; winters from Florida and Texas to the West Indies and Guatemala.

Season—Summer resident in the interior; an irregular summer visitor on Atlantic coast north of the Carolinas; migratory northward to the prairies of the great northwest.

The extraordinary bill of the curlew, curving in the opposite direction from the avocet's, serves the same purpose, however, and drags small crabs and other shell fish that have buried themselves in the wet sand, snails, larvæ, and worms from their holes, the blades acting like a forceps. Beetles, grasshoppers, and flying food seized on the prairies; berries, and particularly dewberries, complete the curlew's menu. The entire bill so far as the nostrils, notwithstanding its extreme length, often sinks through the soft sand or mud to probe for some coveted dainty. The curlew, the avocet, the sea parrot, and the skimmer vie with each other in possessing the queerest freak of a bill.

Large flocks of curlews, flying in wedge-shaped battalions, like geese, with some veteran, a loud, hoarse whistler, in the lead, evidently migrate up our coast to the St. Lawrence and across Canada, to disperse over the broad prairies of the northwest. Not at all dependent on water, however truly their bills indicate that nature intended them for shore birds, they are quite as likely to alight on dry, grassy uplands as on the muddy flats of lower water courses. "Their flight is not rapid, but well sustained, with regular strokes of the wings," says Goss; "and when going a distance, usually high, in a triangular form, uttering now and then their loud, prolonged whistling note, so often heard during the breeding season. Before alighting, they suddenly drop nearly to the ground, then gather, and with a rising sweep, gracefully alight." Flocks on their way south stop to rest awhile on Long Island any time from July to September.

Wherever the curlew strays, its large size and unusual bill make it conspicuous. It is a shy and wary bird, impossible to stalk when feeding, but responsive to an imitation of its call, and coming readily to decoys. In the interior, sportsmen declare the flesh is well worth shooting; but on the coast, north or south, even its odor is rank. Evidently there is a truly strong attachment between members of the same flock, as there is among many sandpipers, for the cries of wounded and dying victims draw the agonized sympathizers back to the spot where they lie, although a second discharge may bring them the same fate.

Three or four clay colored eggs, shaped like a barnyard hen's, but spotted with fine marks of chocolate brown, are found in a depression of the ground. Great numbers of nests are made on the south Atlantic coast and also on the prairies of the northwest, a strange division of habitat indeed for young chicks.

. . . .

Whimbrel, Striped-head, and Crooked-bill, the Hudsonian Short-billed or Jack Curlew *(Numenius hudsonicus)*, with a bill only three or four inches long to bring the entire length of the bird to sixteen or eighteen inches, has blackish brown upper parts mottled with buff, most conspicuous on wing coverts; the

Snipe, Sandpipers, etc.

crown dusky brown, with a buff central stripe; the rest of head, neck, and under parts light buff; a brownish streak running through the eye, and the neck and breast spotted with brown. Flying up the Atlantic coast from Patagonia, the southern limits of its winter quarters, the Jack curlew sometimes loiters awhile in May on our mud flats and marshes before continuing in V-shaped flocks up to the south shore of the St. Lawrence (but not across it), then due north to Hudson Bay, where the nests are built. Evidently nesting duties are soon ended, for returning migrants commonly reach Long Island from July to October. No one has a good reason to give for shooting these birds, yet it is certain that whereas they were once abundant they are now almost rare.

The Eskimo Curlew, Fute, Doe or Dough Bird, Short-billed or Little Curlew *(Numenius borealis);* about thirteen inches long, its short, decurved bill measuring less than two and a half inches, has blackish brown upper parts spotted with buff; the crown streaked, but without the distinct central line that marks the head of the Jack curlew; the under parts buffy or whitish, the breast streaked; the sides and under wing coverts barred with black. *En route* from the Arctic regions, where it nests, to Patagonia, where it winters, this is a very common species at times. The prairie lands adjacent to the Mississippi, its favorite highway, hold "immense flocks" in August and later, it is said; but very few stragglers reach the Atlantic shores. Just as the Jack curlew scrapes acquaintance with the willet, godwit, and other sandpipers on our beaches, so this curlew associates with the upland "plover," the golden plover, and other birds of the interior in this country and on the pampas covered plains of the Argentine Republic. In the Barren Grounds and across the continent from Greenland to Behring Straits, the Eskimo curlew nests. Its whistle is less harsh and loud than its long-billed cousin's, but in their habits generally these three curlews are alike.

PLOVERS
(Family Charadriidæ)

Black-breasted Plover
(Charadrius squatarola)

Called also: BEETLE-HEAD; SWISS PLOVER; BULL-HEAD; WHISTLING FIELD PLOVER; OX-EYE; BLACK-BELLIED PLOVER; MAY COCK

Length—11 to 12 inches.

Male and Female—In summer: Mottled black and white; the upper parts black bordered with white; tail white barred with black; sides of head and neck and under parts black, except lower abdomen and under tail coverts, which are white; axillars (feathers growing from armpits) black. Short bill and the feet and legs black; a small hind toe. *In winter:* Similar, except that upper parts are brownish gray lightly edged with white, and under parts are mixed black and white; but numerous intermediate stages occur, and the plumage is most variable. Immature birds have black upper parts, the head and neck streaked and the back spotted with buff or yellow brown; the breast and sides streaked with brownish gray.

Range—Almost cosmopolitan; nests in Arctic regions, and winters from southern United States to the West Indies and Brazil.

Season—Spring and autumn migrant; May, June; August to October; more abundant in autumn.

Crescent shaped flocks of black-breasted plover, launched on a journey from one end of our continent to the other, come out of the south in May; and following routes through the interior, as well as along the coasts, make short stops only on the way to nest in the Arctic regions. They are now restless, as most birds are in spring. Large and stout for plovers, distinctly black and white while the nesting plumage is worn, there is less danger of confusing them now than in the autumn migration, when immature birds, especially, so closely resemble the golden plover that it is only by noting this bird's small hind toe, which no other plover owns, and the black axillars, or feathers of its armpits, so to speak, where the golden plover is smoky gray, that the sportsman can positively tell which bird he has bagged. It has been said that these plovers migrate in wedge shaped ranks and lines like ducks and geese, which may often be the case, but not always or usual, we think. A cresent shaped flock, the horns pointing sometimes forward, sometimes backward, seems to be the preferred form of flight. Long, perfect wings, a full, slow wing stroke, and a light body are a combination well calculated to discount distance.

Arctic travellers have brought back clutches of three or four pointed eggs that vary greatly in color, ranging between light yellowish olive or dark to shades of brown spotted and speckled with rufous. They have also brought back a "yarn"—or is it a fact?—about the males sitting on the nest and doing all the incubating, while the females enjoy *fin de siècle* emancipation. Fluffy, precocious chicks hatched in June become expert flyers by July, and in August arrive in the United States with parents and friends in motley flocks, often no two birds of which are wearing precisely the same feathers. Having fed chiefly on berries and grasshoppers at the north, autumn birds are counted good eating; but as they have a decided preference for tide water flats and marshes where shrimps and other small marine creatures form their diet, the flesh soon becomes sedgy and unpalatable once they reach the coast. A quick strike at a particle of food about to be picked up makes these plovers appear greedy; however, all their motions are quick and sudden. In running, especially, are they nimble: a sprint of a few yards, a sudden halt to reconnoitre with upstretched heads, another quick run, then a pause, are characteristic movements of most plovers, just as squatting to conceal themselves is.

Because so many young innocents which have no knowledge of men make up the autumn flocks, these respond quickly to decoys and to an imitation of their clear, mellow whistle, that penetrates to surprising distances from where the birds are circling high in air. Down they sail on motionless wings, apparently glad for any diversion in their aimless, roving life. Soft notes of contentment uttered as they drift with decurved wings and dangling legs toward the decoys are soon silenced forever by a deadly report. Twenty years ago the black-breasted and the golden plovers were abundant on the Iowa and Illinois prairies in spring and fall, but they were pursued by sportsmen so relentlessly that now a flock is seldom seen in either state. The few birds that remain seem to have chosen some other route for their migrations, in order to escape the fusilades to which they were there subjected.

American Golden Plover

(Charadrius dominicus)

Called also: FIELD PLOVER; GREENBACK; GREEN PLOVER; PALE-BREAST; TOAD-HEAD; PRAIRIE PIGEON; FROST BIRD; SQUEALER

Length—10.50 inches.

Male and Female—In summer: Mottled upper parts black, greenish, golden yellow, and a little white, the yellow in excess; tail brownish gray indistinctly barred with whitish; sides of breast white; other under parts and sides of head black; under wing coverts ashy gray. Bill and feet black. *In winter:* Upper parts and tail dusky, spotted or barred with yellow or whitish, the colors not so pure as in summer; under parts grayish white, purest on chin and abdomen; the throat and sides of head streaked; the breast and sides of neck and body mottled with dusky grayish brown; legs dusky. Immature birds resemble winter adults; also like black-breasted plovers; but the grayish axillars and the lack of a fourth toe sufficiently distinguish this species from the preceding, however variable the plumage may be at different seasons.

Range—North America at large; nests in Arctic regions; winters from Florida to Patagonia.

Season—Spring and autumn migrant; May; August to November.

Golden grain, golden rod, golden maple leaves, and golden plover all come together; the birds not so yellow, it is true, as they were in the spring, when they gave us only a passing glimpse of their clearer, more intense speckled plumage, but still yellow enough to be in harmony with nature's autumnal color scheme. Indeed, they blend so well with their surroundings as to be all but invisible. Usually the under parts of birds are light colored to help make them inconspicuous on the wing; but the black markings on this and the preceding plovers are notable exceptions. High above the corn and buckwheat, the stubble, the burned and ploughed fields of the interior, or the level stretches of grass far back of the beaches, the sandy dunes, and flats bared at low tide along the coast, come the plovers in crescent-shaped flocks, now massed, now scattered, now rising, now dipping, the wings tremulous with speed, and swinging round in a circle at sight of a feeding ground to their liking. With soft, trilled mellow whistles rippling from their throats, the birds drift downward on set, decurved wings, and skim low before alighting. For an instant, as their dangling legs touch the ground, they raise their wings high above their backs until they meet, then slowly fold them against their sides. Now they scatter, and running nimbly and gracefully hither and thither, check themselves suddenly from time to time, raise their heads and look about to reconnoitre. Every motion is quick; they strike at a particle of food as if about to take a dive loon fashion, then run lightly on again, soon returning to the same spot if driven off. A hasty run must be taken, even when frightened, before the plovers spring into the air. A flock has a curious way of standing stock still at an alarming noise, before starting to run. When they squat and hide behind tufts of beach grass, it takes sharp eyes to detect birds from sand.

But even without apparent alarm, the scattered birds often rise as if summoned by some invisible and inaudible captain, and fly close along the ground, wheeling and dashing and skimming in beautiful and intricate evolutions. Such a flock offers all too easy a side shot. In "the good old days" of carnage that are responsible for the scarcity of this fine game bird to-day, it often rained plover when the gunners were abroad. This latter phrase suggests the query: What connection of ideas is there between *pluvia* (rain) and plover derived from that word? An early French writer, Belon (1555), speaking of the European species, of course, says "Pour ce qu'on le prend mieux en temps plurieux qu'en nulle autre saison;" but with us the birds are, if anything, wilder and less approachable in rainy weather than when it is fine. Is it that their backs look as if they had been sprinkled

AMERICAN GOLDEN PLOVER.
½ Life-size.

with rain drops ; or that they whistle more before storms, as their German name (Regempfeifer) would imply; or that the east wind that brings rain, blows flocks of these migrants in from sea ?

Golden plovers, once so plentiful and confiding that they came near enough to the plough for the farmer's boy to strike and kill with his whip, were sold in the Chicago streets for fifty cents a hundred within the memory of many, and those not the oldest inhabitants. Dead birds propped up with sticks when the wooden decoys from city shops were not available ; a dried pea rattling about in a hollow reed to imitate the mellow *coodle, coodle, coodle* of the plover's melodious call, allured the birds within easy range of every farm hand's antediluvian musket.

Plovers' visits depend much on weather, a clear, fine day inviting a long, unbroken flight far out at sea during the autumn migration ; whereas lowering weather, especially an easterly storm, drives the birds to the coast, where, flying low, a warm reception of hot shot usually awaits them from behind blinds. Grassy level stretches and pasture lands back of the beaches, rather than sandy places, attract them, since land insects, grasshoppers particularly, and worms are what they are ever seeking. In the autumn migration, at least, the great majority of plovers follow the coast, sometimes closely, sometimes far at sea, so far that many flocks on their way to South America pass to the east of Bermuda. Long, perfect wings and light bodies enable them to cover immense distances without resting. While no fixed route appears to be followed in spring, possibly the birds show a preference then for the freshly-ploughed inland fields where food, winged, crawling, and in the larval state, abounds.

Among all the gaily dressed, tuneful lovers that visit us in May, few are handsomer and more charming in voice and manner than this melodious whistler. Further north he breaks into a long serenade, sung chiefly in the short Arctic night : *tee-lee-lee, tu-lee-lee wit, wit wit, wee-u-wit, chee-lee-u-too-lee-ee*, as described by Wilson, who followed these plovers to Behring Sea until he found their nest, that so few know. A depression among the grass or moss, lined with fine grasses and dried leaves, usually cradles four yellowish eggs covered over with dark reddish brown spots ; but in the eggs, as in the plumage of the plovers, there is great variation. Birds that lay pointed eggs, as plovers do, arrange the narrow ends toward the centre of the nest that they may be the better covered; and rumor says these emancipated females leave all the incubating to the males.

Killdeer

(Ægialitis vocifera)

Called also: KILLDEE; KILDEER PLOVER

Length—9.50 to 10.50 inches. About the size of the robin.

Male and Female—Grayish brown washed with olive above; the forehead, spot behind eyes, throat, a ring around the neck, a patch on wing, a band across breast, and underneath, white; front of crown, cheeks, a ring around neck, and a band across breast, black; lower back and base of tail chestnut; inner tail feathers like upper parts; outer feathers chestnut and white, all with subterminal band of black tipped with white. Bill black; legs light; eyelids red.

Range—Temperate North America to Newfoundland and Manitoba; nests throughout range; winters usually south of New England to Bermuda, the West Indies, Central and South America.

Season—Resident, March to November, or later; most abundant in spring and autumn migrations.

A certain corn field used to be visited daily by an aspiring ornithologist, aged nine, for a peep at four little yellowish white eggs, spotted and scrawled with chocolate brown, that were laid directly on the ground, without so much as a blade of grass to cradle them. Every visit threw the old birds into a panic, which, of course, was part of the fun anticipated in making the visit. *Kildeer, killdeer, dee, dee,* they called incessantly as they whirled about overhead and screamed in the child's ears; but still the eggs were relentlessly fondled, while the mother now frantically ran about, dragging her wings beside her, pretending to be lame; now sprang into the air and dashed about every which way, as if mad. In spite of much handling, however, the eggs actually hatched; and what was the child's amazement after leaving them at nine o'clock one morning to return at ten and find eggs, birds, and even shells had disappeared! Later a brood of queer, top-heavy, long-legged, striped, and downy chicks was discovered running nimbly about the corn field, feeding; but what they did with their eggshells ever remained a mystery.

This common plover of pastures and cultivated fields, of lakesides and marshes, or any broad tracts of land near water, that seems indispensable to its happiness, is in decided evidence because of its wild, noisy cry even when we cannot see the bird; but the two black bands across its breast, its white forehead and red eyelids easily identify it whenever met. As a rule one sees flocks of these plovers only a-wing, for they scatter when feeding. Sometimes the *kill-dee, kill-dee* sounds low and sweet, with a plaintive strain in it; but let any one approach the bird's haunts, and the voice rises higher and shriller until it would seem the strident notes must soon snap the vocal cords. Cows, horses, sheep, and the larger poultry that wander over a farm do not alarm these birds in the least. In their presence they are gentle and almost tame, but a man is their abhorrence in regions where they have been persecuted; elsewhere they are not conspicuously wild. Yet their flesh is musky and worthless from the point of view of the sportsman, who seldom wastes shot on it. A startled bird will run swiftly away rather than fly at first, stop occasionally to look back at the villain still pursuing it, crying complainingly all the while, and perhaps flutter in low, short flights to lure the intruder still farther away. But the killdeer, with its long, perfect wings, is a strong, steady high-flyer, however erratic and uncertain its flight may be when suddenly flushed by some innocent stroller taking a short cut through the pasture. Restless and full of fears, real or imaginary, there is scarcely an hour of the day or the night when its voice is not raised, until sportsmen have come to regard so keen a sentinel as a nuisance. Dr. Livingston met with a close kinsman of the killdeer in Africa that he described as "a most plaguey sort of public spirited individual that follows you everywhere, flying overhead, and is most persevering in his attempts to give fair warning to all animals within hearing to flee from the approach of danger."

On the ground, where the killdeer spends most of its time, it moves about daintily, quickly, even nervously; for it never remains still except for the instant when it seems to gaze at an intruder with withering contempt. Since worms, that are its favorite food, come to the surface after sundown, this bird, like many others of similar tastes, is partly nocturnal in habits; but grasshoppers, crickets, and other insects take it abroad much by day. It migrates chiefly at night, the *killdeer, killdeer*, resounding from the very stars.

Semipalmated Plover

(Ægialitis semipalmata)

Called also : RING-NECKED OR RING PLOVER

Length—6.75 to 7 inches. A trifle larger than the English sparrow.

Male and Female—Upper parts brownish gray; front of crown, band across base of bill, sides of head below eye, and band on breast, that almost encircles the neck, black; forehead, throat, ring around neck, parts of outer tail feathers, and under parts white. Brownish gray replaces the black in winter plumage. Bill black, orange at base; ring around eye bright orange; yellow toes, webbed at the base.

Range—North America at large; nesting from Labrador and Alaska northward to Arctic sea; winters from Gulf states to Bermuda, West Indies, Peru, and Brazil.

Season—Spring and autumn migrant; April, May; July, August, September; most plentiful in late summer and early autumn.

Closely associated with the friendly little sandpipers, these small plovers likewise haunt the beaches, their plumage in autumn being precisely the color of the wet sand they constantly run about on in small scattered flocks. When the tide goes out, their activities increase. Birds that have been hiding in the marshes and sand dunes now trip a light measure over the exposed sand bars and mud flats, leaving little tracks that may not be distinguished from those of the sand ox-eye or semipalmated sandpiper that hunts with them, although the plover has only three half webbed toes. The small, slightly elevated fourth toe of the ox-eye is only faintly evident at times in its tracks.

Tiny forms chase out after the receding waves, running in just in advance of the frothing ripples that do not quite overtake them, although the plovers almost never spring to wing as sandpipers do when a drenching threatens, but place all their trust in their fleet legs. With such feet as theirs, they must be able to swim; but who ever sees them in deep water? More silent, too,

RING PLOVER.
Life size.

than sandpipers, it is chiefly when alarmed that two plaintive, sweet, but sometimes sharp notes escape them, whereas sandpipers keep up their cheerful *peep, peep,* under all circumstances. Real danger summons the scattered flocks of ring-necks to wing into a compact mass that moves as if swayed by one mind; but like most birds that nest too far north to become acquainted with murderous men, these gentle, confiding little plovers suspect no evil intentions and rarely fly away. Running to hide by squatting behind tufts of beach grass stills their small fears.

In the interior, for an inland route is followed as well as a coastwise one, the ring-neck runs about the margins of small lakes or ponds, rivers and marshes, everywhere looking for worms, small bits of shell fish, eggs of fish, and insects; always alert and busy and hungry. General Greeley found these plovers still nesting in Grinnell Land early in July; yet by the end of the month stragglers from large flocks begin to arrive in the United States—a little journey to try the wings of fledgelings *en route* to Brazil. It is said the male arranges the small pear shaped buff eggs, spotted with chocolate, with the pointed end toward the centre of the depression in the ground that answers as nest, the better to cover all four with his breast, for it is he who does most, if not all, of the incubating. Greenlanders, who have a longer opportunity to study this interesting little bird, say that it claps its wings before a storm and becomes strangely excited; but although it has the dainty habit of lifting its wings high above its back till they meet, on alighting, no excited clapping of them has been recorded here. This is the most abundant and most widely distributed of the ring-necks.

Piping Plover
(Ægialitis meloda)

Called also : PALE RING-NECK

Length—7 inches. A trifle larger than the English sparrow.

Male and Female—Upper parts very pale ash; forehead, ring around neck, and under parts white; front of the crown and a link of incomplete collar either side of breast, black; inner tail feathers dusky, the outer ones becoming white. In winter plumage the black is replaced by brownish ash.

Plovers

Range—North America east of the Rockies; nesting from coast of Virginia northward to Newfoundland; winters in West Indies.

Season—Summer resident, March to September; most abundant in autumn migrations.

Very slight differences in the habits of plovers that haunt our beaches have been noted by the most tireless students, and were it not for the piping plover's notes there would be nothing beyond a reference to its stronger maritime preferences and more southerly nesting range to add to the account of the ring-neck. The piper, much lighter in color, is the lightest species that visits us. It nests among the shingle on our beaches from Virginia to Maine and beyond, where it is next to impossible to discover the finely speckled drab eggs that imitate the sand perfectly; and possibly because it does not pass half its year in Arctic seclusion, as some other plovers do, it is not quite so gentle and confiding as they—this is the sum of its peculiarities. Its pathetically small size, scarcely larger than that of an English sparrow, should be, but is not, a sufficient protection from the gun.

"It cannot be called a 'whistler' nor even a 'piper' in an ordinary sense," says Mr. Langille. "Its tone has a particularly striking and musical quality. *Queep, queep, queep-o*, or *peep, peep, peep-lo*, each syllable being uttered with a separate, distinct, and somewhat long drawn enunciation, may imitate its peculiar melody, the tone of which is round, full, and sweet, reminding one of a high key on an Italian hand-organ or the *hautboy* in a church organ." The sweet, low notes, it should be added, have an almost ventriloqual quality also, that often makes it difficult to locate the bird by the ear alone.

Retiring to the dunes and meadows back of the beach only to sleep or rest when the tide is high, we most frequently see this active little sprite running nimbly along the wet sands, poking among the shells, chasing out after the waves, and hurriedly picking up bits of food before being chased in by them, or flying above the crests short distances along the beach, usually to escape a deluge from the combing breakers. All its movements are alert, quick, graceful. At Muskegat, where this plover's nests are found among the terns', the plover loses little by comparison with those preëminently graceful birds. Around the great lakes scattered flocks are seen in the migrations chiefly; but it is on the secluded Atlantic beaches, comfortably distant from seaside resorts, that we find the piping plover most abundant.

The Belted Piping Plover (*Ægialitis meloda circumcincta*), a western representative of the preceding, differs from it only in having the black links on the breast joined to form a band.

The Mountain Plover (*Ægialitis montana*), a distinctly prairie bird, rather than a mountaineer, has grayish brown upper parts, the feathers margined with chestnut; the white under parts grow yellowish on breast, but without belt or patches; the front of the crown and the cheeks black. It is almost nine inches long. It has all the charming grace, quickness of motion, and winning confidence that characterize its clan.

Wilson's Plover
(*Ægialitis wilsonia*)

Length—7.50 inches.

Male and Female—Upper parts ashy brown, tinged on nape and sides of head with chestnut; forehead and under parts white, the white of throat passing around like a collar, and the white of forehead running backward in a line over each eye to nape; lores, front of crown, and a band across the breast black in male, brownish gray in female; inner tail feathers dark olive, the outer ones becoming white. Bill large, stout, and black, no colored eye ring; legs flesh colored. Immature birds look like mother, but have upper parts margined with gray or white, more closely resembling dry sand, if possible, than do the adults.

Range—America, nesting from Virginia southward; winters south to Central and South America; common on south Atlantic and Gulf beaches and California.

Season—Summer resident; a few winter in the south.

A beach bird in the strictest sense, Wilson's plover is never found inland, but close beside tide water on the mud flats that furnish a fresh *menu* at each ebbing; or on the dry sand beyond the reach of the surf, where its plumage, in wonderful mimicry of its surroundings, conceals it perfectly. In the short, sparse grass

Plovers

of the upper beaches, a brooding bird that knows enough to keep still in the presence of a passer-by, runs little risk of detection. The three clay colored eggs, evenly and rather finely spotted and speckled with brown, that are laid directly on the sand, require little incubating, however, beyond what the sunshine gives them; but the parents never stroll so far away from their treasures that they may not return instantly danger threatens and run or swoop about the visitor, imploring retreat. Gentle, unsuspicious manners give these birds half their charm. Their grace of motion, another characteristic, suffers little by comparison with that of the terns not infrequently found nesting among them. On the ground all plovers excel in sprightliness; every movement is quick and free; and on the wing, also, these describe all manner of exquisite evolutions, half turning in the air to show now the upper, now the under side of the bodies; now sailing on long, decurved, motionless wings; now hovering an instant before alighting, stretching their wing tips high above the back—a beautiful posture that the terns have evidently copied.

Quite closely resembling the semipalmated plover in plumage, this species may always be known by its large, heavy bill, the largest, in proportion to the size of the bird, any plover has, and by the absence of a bright eye ring that, with the partial webbing of its toes, are the ring-neck's diagnostic features. Small flocks of Wilson's plover reach Long Island every summer, but rarely touch the New England coast. The morsel of flesh on its plump little breast should seem not worth the hunting by healthy men, whose appetites need no coaxing. One who little understands the ways of gunners might think a bird smaller than a robin would suffer little persecution.

Dr. Coues describes this plover's note as half a whistle, half a chirp, quite different from the other plovers' calls; but a plaintive quality can be detected in it, too, as in the voices of most beach birds, that reflect something of the mystery and sadness of the sea. In his lines to "The Little Beach Bird," that are applicable to a dozen species, Richard Henry Dana emphasizes the contrast between the joyous songs of land birds and the melancholy, plaintive strains of those that live along the sea.

SURF BIRDS AND TURNSTONES
(Family Aphrizidæ)

Turnstone
(Arenaria interpres)

Called also:—BRANT BIRD ; CALICO BACK ; CHECKERED SNIPE ; HORSEFOOT SNIPE ; HEART BIRD ; CHICKEN PLOVER ; RED-LEGGED PLOVER

Length—8.50 to 9.50 inches. A little smaller than a robin.

Male and Female—*In summer*: Upper parts strangely variegated and patched with chestnut, black, brown, and white; base of tail white, the tail feathers banded with black and tipped with white; white band on wings; beneath, including wing linings, white; the throat and breast jet black divided by a white space. Black, short bill tapers to an acute tip, very slightly recurved; legs orange red; the small hind toe turns inward. The female has less chestnut and more plain brown on her upper parts, and the black lacks the lustre of jet. *In winter:* Upper parts blackish blotched with gray and brown or ashy brown, and lacking the chestnut feathers; the breast markings more restricted.

Range—Nearly cosmopolitan; nests in Arctic latitudes and in the Western hemisphere; migrates to South America so far as Patagonia.

Season—Irregular, **transient visitor;** April, May; August, September, or later.

With a bill curiously like a writing pen, this well named wader turns over pebbles, clods of mud, shells, and seaweed on the beaches more commonly about the foot of cliffs and in stony coves than on long, sandy stretches, ever looking for the small marine creatures that satisfy its appetite, particularly for the eggs of the horsefoot or king crab (*Limulus polyphemus*), its favorite dainty. Often not only the head and bill must be used to push over a stone, but the breast assists too; ordinarily, however, the bird

Turnstones

simply pokes its bill under a lighter object, and, giving its head a quick jerk, turns over the roof under which some small prey thought itself secure, swallows the morsel, then runs off to the next shell to repeat the operation. Seaweed is simply tossed aside.

Joseph's coat doubtless showed no more variegated patchwork than the turnstone's nesting plumage, which, however, differs greatly in individuals, scarcely any two of which have precisely the same markings at any season. Because of this variety the early ornithologists believed there were several more distinct species of turnstones than actually exist. Other beach birds are mostly clad in soft tints that so blend with the sand we can scarcely distinguish them until they move; but the calico back, although small, is ever conspicuous, and possibly because it knows how hopeless concealment is, as compared with the confiding, gentle little sandpipers and plovers, it is shy and wild.

Small companies of three or four, or family parties, run about the outer beaches with all the sprightliness of plovers, then stop suddenly to meditate, then run on again, pausing to turn over a shell now and then, but always active, and more ready to place dependence on their fleet legs than on their wings to distance a pursuer; yet when one goes too near, the turnstone rises, uttering a few twittering, complaining notes, flaps its wings quickly, sails low, and with a few more flaps and another sail soon alights at no great distance, to return to the point where it was flushed at its first opportunity. It is wonderfully patient and persistent about exhausting the resources of one feeding ground before looking for another. Wading about in a cove, it will sometimes deliberately seat itself in the water, just as it squats on a beach, and swim off easily to a safe distance across the inlet from the intruder.

A bird that travels from Patagonia to the Arctic Circle to nest, naturally is a fast, strong flyer, the frequent sailings after quick flaps of the wings resting them sufficiently to make long, uninterrupted flights possible. General Greeley found turnstones as far north as he went, and reported that fledgelings which in late June had emerged from clay colored eggs (blotched and scrambled with grayish brown) were able to fly by the ninth of July. A few birds take an inland route during the migrations, and display their freaky feathers on the shores of the Great Lakes and larger rivers.

OYSTER-CATCHERS
(Family Hæmatopodidæ)

American Oyster-Catcher
(Hæmatopus palliatus)

Called also: BROWN-BACKED OYSTER-CATCHER; FLOOD GULL

Length—17 to 21 inches.

Male and Female—Head, neck, and upper breast black; back and wings dark olive brown; greater wing coverts, base of secondaries, sides of lower back, upper tail coverts, base of tail, and all under parts, white. Bill coral red, twice as long as head, compressed, almost like a knife-blade at end, but varying in shape, owing to wear and tear; feet flesh colored; three toes united by a membrane to middle joint.

Range—" Sea-coasts of temperate and tropical America, from New Jersey and Western Mexico to Patagonia; occasional or accidental on the Atlantic coast north to Massachusetts and Grand Menan." A. O. U.

Children brought up on " Alice in Wonderland " might imagine from the name of this bird that oysters are fleet-footed racers along our beaches, overtaken at the end of a breathless chase by the oyster-catcher! On the New Jersey coast and southward, but rarely farther north, we see (if we are cautious, far sighted-stalkers), this curious bird actually prying open shells of bivalves—oysters less commonly, however, than mussels and some others—and digging up fiddler crabs and worms that have buried themselves in the soft sand, with a bill that is one of the most peculiar among bird tools. Long, stout at the base, but compressed like a knife blade at the end; often as worn and jagged as the opener seen at a Coney Island oyster stand; sometimes bent sideways from severe wrenches, and bright coral red—this bill belongs in the same class of freaks as the bills of the avocet, skimmer, curlew,

Oyster-catchers

woodcock, and sea parrot. The oyster-catcher is a shy bird, constantly on the alert, and it is no easy matter to steal upon one close enough to watch it at work. Walking with stately dignity along the lower beach, striking its bill into the sand, often up to the nostrils, suddenly it stops at a glimpse of an intruder, and with shrill notes of alarm springs into the air and is off, not in a short flight, as the confiding little plovers and sandpipers make, soon to return, but away down the beach, often out of sight. Another time you will have learned to rely on a powerful field glass to lessen the distance between you.

But this bird, so quick to move out of harm's way, is a past master in the art of stealing upon bivalves unawares when they are lying about on the beaches with their shells open, and prying the shells apart until the delectable morsels are cut from them and swallowed. Whoever has had his finger pinched between mussel shells will not be surprised at the crooked, jagged blade the oyster-catcher often carries about. When the bird finds its bill hopelessly caught in a vise, it simply lifts the razor clam, "racoon oyster," or whatever its captor may be, knocks it against a rock until the shell is broken, and then feasts. Limpets are pried off rocks as if with a chisel. Again the oyster-catcher wades into the shallows for shrimps and other little marine creatures. No doubt it can swim well too, owing to the partial webbing of its toes, but rapid running and still more rapid flying usually make other accomplishments superfluous. With tough, unsavory flesh to save it from sportsmen's persecutions, it is a timid bird, nevertheless. It does not live in large flocks; solitary, or with two or three companions only, it dwells far from the haunts of men and apart from those sociable beach birds that are too confiding for self-preservation. A striking, handsome wader on the ground, it is even more attractive as it flies with a few friends, showing its glistening white under parts as it wheels about overhead with great regularity of manœuvre. Rapid wing beats and frequent sails make its flight strong, yet extremely graceful. A quick, shrill *wheep, wheep, wheo*, uttered on the wing as well as on the ground, voices the bird's various emotions. Birds of a migrating flock are said to keep together in lines like a marshalled troop, swayed by one mind, just as they appear to be when wheeling over the beach on pleasure bent.

Like gulls, terns, skimmers, and other beach nesters, the oyster-catchers allow the sun-baked sand to do the greater part of the incubating, the parents confining themselves only at night or during storms on three or four pale buff eggs spotted and blotched with chocolate, and laid directly on the shingle, in a depression. Mr. Walter **Hoxie, in the** " Ornithologist and Oölogist," tells of seeing **a pair of these** birds whose nest had been discovered, but not **disturbed,** take the eggs about one hundred yards farther **along the** beach and deposit them safely, one by one, in a new **nest** which he watched them prepare. Fluffy chicks, that run as soon as hatched, will squat and remain motionless like plovers, secure in their plumage's perfect imitation of their surroundings.

PART III

GALLINACEOUS GAME BIRDS

GALLINACEOUS GAME BIRDS

Bob Whites
Grouse
Turkeys

GALLINACEOUS GAME BIRDS
(Order Gallinæ)

Birds that scratch the ground for food, the progenitors of our barn-yard fowls, the game birds *par excellence* of the sportsman, none are more interesting either from his point of view or from that of the bird student, or of greater commercial value. Certain structural peculiarities are noticeable throughout the group: a greatly enlarged **esophagus**, now called a crop, receives the bolted food and moistens it, leaving to a very thick, hard gizzard (except in the sage cock) the work of grinding the food with the help of gravel swallowed with it. Usually heavy in body, round breasted, small of head, stout of legs and feet, sometimes with spurs on the former, richly, if often quietly, plumed, the appearance of these birds is too familiar to be enlarged upon. They are prolific layers, and raise large broods, that follow the mother like chickens, as soon as hatched, one or more families composing a covey or bevy soon after the nesting season.

Bob Whites, Grouse, etc.
(Family Tetraonidæ)

Of the two hundred species contained in this great family, one-half belong to the Old World, where they are known as partridges and quail, names miscellaneously applied to our grouse and Bob Whites, that differ greatly in structure from their European allies, and the source of endless confusion in the popular mind. Three subfamilies go to make up this large family: the *Perdicinæ*, or Old World partridges and quail; the *Odontophorinæ*, or New World partridges and Bob Whites; and the *Tetraoninæ*, or grouse. These fowl-footed birds have the hind toe raised above the ground, differing from the pigeon-footed gallinaceous birds, that have four toes on the same level;

and the grouse have feathered legs, like many birds of prey, to keep these parts from being frozen, since they frequent high altitudes. None of these American species is migratory, yet their rapid, *whirring* flight, performed with quick strokes of small, concave, stiffened wings, is well sustained, and sometimes for long distances. The heads of grouse especially, high at the rear to contain the unusually developed brain, indicate that rare degree of intelligence among birds which so taxes the wits of the sportsman; but certainly the Bob White is not lacking in mental calibre. The latter birds are devoted lovers and parents, whereas grouse are generally polygamous, and the males are either indifferent to the eggs and young, or, in some cases, destructive of them. Mr. D. G. Elliot remarks: "It is a rather singular fact that in most polygamous species the plumage of the sexes is very dissimilar, while there is usually but little difference observable between those that are monogamous."

> Bob White, or Quail.
> Dusky, or Blue Grouse.
> Canada Grouse, or Spruce Partridge.
> Ruffed Grouse, or Partridge.
> Canadian Ruffed Grouse.
> Gray Ruffed Grouse.
> Oregon, or Red Ruffed Grouse.
> Prairie Chicken, or Pinnated Grouse.
> Prairie Sharp-tailed Grouse.
> Columbian Sharp-tailed Grouse.
> Sharp-tailed Grouse.
> Sage Grouse, or Cock of the Plains.

Pheasants and Turkeys

(Family Phasianidæ)

A group of magnificent birds, including the peacock, pheasants, and the jungle fowl, the progenitors of our domestic poultry. From the Mexican turkey, now imported all over the world, and into France and England since the sixteenth century, came the race that furnishes our Thanksgiving feasts.

> Wild Turkey.

BOB-WHITE.
⅓ Life-size.

BOB WHITES, GROUSE, ETC.

(Family Tetraonidæ)

Bob White

(Colinus virginianus)

Called also: QUAIL; PARTRIDGE; VIRGINIA PARTRIDGE

Length—9.50 to 10.50 inches.

Male and Female—Upper parts reddish brown or chestnut, flecked with black, white, and tawny; rump grayish brown, finely mottled, and with a few streaks of blackish; tail ashy, the inner feathers mottled with buff; front of crown, a line from bill beneath the eye, and band on upper breast, black; forehead, and stripe over the eye, extending down the side of the neck, white; breast and under parts white or buff, crossed with irregular narrow black lines; feathers on sides and flanks chestnut, with white edges barred with black. The female has forehead, line over the eye, and throat, buff, and little or no black on upper breast. Summer birds have blacker crowns and paler buff markings. Much individual variation in plumage.

Range—" **Eastern** United States and southern Ontario, from southern Maine to the south Atlantic and Gulf states; west to central South Dakota, Nebraska; Kansas, Oklahoma and eastern Texas. Of late years has gradually extended its range westward along lines of railroad and settlements; also introduced at various points in Colorado, New Mexico, Utah, Idaho, California, Oregon, **and** Washington. Breeds throughout its range." A. O. U.

Season—Permanent resident.

Endless confusion has arisen through the incorrect local names given to the Bob White, which in New England is called quail wherever the ruffed grouse is called partridge, and called partridge in the middle and southern states wherever the ruffed grouse is called pheasant; but true partridges and quail, quite different in habits and appearance from ours, are confined to the Old World, however firmly their names cling to the American species. That which we call a quail, by any other name would taste as sweet; and it is surely time the characteristic game bird of this country received in all sections its characteristic, distinctive title. Bob White, the name it calls itself, also has the sanction of that dignified, conservative body, the American Ornithologists' Union, than which can there be two higher authorities?

Before the snow and ice have been melted by spring sunshine, *Bob White! ah, Bob White!* a clear staccato whistle, rings out from some plump little feathered breast swelling with tender and sincere emotions. Mates are not easily won: sharp contests of rival males, that fight desperately, like game cocks, occur throughout the pairing season; the demure, coy little sweetheart, concealing her admiration for the proud victor strutting before her, only fans his flame by her feigned indifference. In vain he jumps upon a stump and, like a ruffled orator, repeats his protestations. He runs beside her, now bowing, now crossing her path, ardently entreating some sign that his handsome feathers, his gallantry, his musical voice, his sworn devotion to her, have made an impression; but the shy little lady, appearing to be frightened by such ardor, discreetly withdraws, knowing perfectly well, as every coquette must, that such coyness never discourages a suitor worth the having. Marriage is not entered into lightly or irreverently by these monogamous birds, unlike their European Mormon kin that utterly lack the gallantry and affectionate nature characteristic of the American bird. It is a slander to call Bob White by the name of the disreputable, pugnacious, selfish, mean-looking quail. Rarely, indeed, does he lapse from rectitude and take a second mate.

In May, a simple nest, or slight depression in the ground, lined with leaves and grasses, is formed sometimes in the stubble, in a grassy tussock that meets overhead, and must be entered from one side; or beneath a small bush, next a worm-eaten old log, at the foot of a stump; or in the cotton rows—anywhere, in fact, where seclusion favors. Some nests have been found with well constructed domes, and the entrance a foot or more from the nest proper. Incredibly large numbers of brilliant white eggs—as many as thirty-two—are reported in a single nest, all skilfully packed in, pointed end downwards to economize space. Does the amiability of the female extend to sharing her nest with a rival, or are all these eggs hers? Remove an egg, and it is impossible for the human hand to rearrange the clutch with such faultless economy. In the middle and southern states, where two and even three broods have been reared in a season, the number of eggs laid at a time rarely exceeds ten, so that the autumn coveys there are no larger than those in the north. Both parents take turns in covering the eggs, the male encouraging his brooding mate by cheerful, musical whistles introduced by a half-suppressed syllable, that the New Englanders translate into *No more wet! more wet!* or *Pease most ripe! most ripe!* and the Western farmers into *Sow more wheat! more wheat!* A shrill *wee-teeh*, used as a note of warning; *quoi-hee, quoi-hee*, to reassemble a scattered covey; a subdued clucking when undisturbed, and a rapidly repeated twitter when surprised, are Bob White's vocal expressions. One feels happier for having heard his exuberant joy and pride whistled from a fence-rail or low branch of a tree. How readily he answers the farmer's boy whistling to him from the plough! He is decidedly in evidence, bold and fearless during the twenty-four days of incubation; but one rarely sees the female then. She is ever shy. Ray, the English naturalist, says the European quail hatches one-third more males than females—a proportion that corresponds with the numbers generally bagged by our gunners. Should the eggs be handled when first laid, the nest is at once deserted. Mowing machines work sad havoc every year.

Precisely as a brood of chickens follows a mother hen about the farm, so a bevy of comical little downy Bob Whites, sometimes with the shells still sticking to their backs, run about through the tangled brake and cultivated fields, learning from both devoted parents which seeds of grasses, cereals, and berries they may eat. Farmers bless them for the number of weed-seeds and insects they destroy. The fox and the hawk, next to man, are their worst enemies. A note of alarm sends the fledgelings half-running, half-flying, to huddle up close to the mother; or when a cold wind blows, a soft, low, caressing twitter summons the babies to shelter beneath her short wings, that barely cover the large brood.

Later, the young scatter and hide among the grass, while the parents, feigning lameness and the usual pathetic artifices familiar to one who has ever disturbed a family of young birds, dare all things for their dear sakes. Should some accident befall the female during incubation, the male faithfully covers the eggs and ministers to every want of his happily precocious family; and in the south, where the female frequently begins to lay again when her first brood is but a few weeks old, it is the father, a pattern of all domestic virtues, that then assumes its full care. When the second brood leaves the shell, one large happy family, known in sportsman's parlance as a bevy or covey, makes as charming a picture as one is likely to meet in a year's tramp. Southern sportsmen, especially, sometimes express surprise at finding birds still in pin feathers and unable to fly in November, when part of the brood, at least, may not be distinguished from adults; but these most prolific of all game birds not infrequently devote six months to nursery duties. Bob Whites are eminently affectionate, and a covey never willingly disperses until the spring pairing season.

"It is a glorious day: come, let us kill something!" says London *Punch's* famous sportsman; and when the splendor of autumn glorifies our fields and woods, domed by a sky of clearest, most intense blue, and the keen, frosty, sparkling air invigorates both mind and body, the American sportsman likewise takes down his light, short gun and some shells loaded with No. 8 shot, whistles up his dog, which nearly twists himself inside out with happiness, and at sunrise is off. Now the coveys are feeding in the field of buckwheat—a favorite resort—or in the stubble of the corn, rye, or oat fields, or along the ditches and clearings fringed with undergrowth, or in the vineyard or orchard—just where it is the dog's business, not the author's, to disclose. The seed of the locust, wild pease, tick, trefoil, sunflower, smartweed, partridge berry, wintergreen and nanny berries, acorns, and beechnuts do not complete the Bob Whites' *menu*. Late in the forenoon, the hearty breakfast having at length ended, a bevy of birds will first slake their thirst before huddling together to preen and dust their feathers and enjoy a midday siesta on a sunny slope. They keep near water during droughts; but after long rains, look for them on the dry uplands and along the sunniest coverts, not too early on a frosty morning, when they are likely to remain huddled together late to keep warm until the hoar frost melts in the sunshine. These birds have a unique manner of sleeping: forming a circle on the ground, in a sheltered open, beyond thickets where prowling fox and weasel lurk, they squat close together as they can huddle to save heat, and with their tails toward the centre, and their heads pointing outward to detect danger from every possible direction, rest secure through the night and sometimes part of cold and stormy days, the male parent usually remaining outside the ring to act as sentinel. As winter approaches, they leave the open, cultivated fields to withdraw into sheltered thickets and bottom lands, sometimes to alder swamps. Now, when hunger often pinches cruelly, the food scattered for barnyard fowls is fearlessly picked up; indeed, these birds haunt the outskirts of farms at all seasons, following the pioneer and railroad westward, and ever going more than half way in establishing friendly relations between themselves and mankind. While all efforts to domesticate them have ended in runaways when the nesting season came around and wild birds whistled enticing notes of happiness and freedom, protection from the shooters, and a few handfuls of buckwheat scattered about for them in the bitter weather are all the encouragement these appreciative little neighbors need to keep them about the farm. Like the ruffed grouse they will allow the snow to bury them, or voluntarily bury themselves in it to escape extreme cold; but an ice crust forming over a sleeping covey often imprisons it, alas! and not until a thaw is the tragedy revealed in a circle of feathered skeletons.

A loud *whir-r-r-r-r-r*, as a flushed flock rises to wing, indicates something of the speed at which the Bob Whites rush through the air. They are not migratory, usually remaining resident wherever found, although from the northern boundary of their range coveys seen travelling afoot in autumn certainly appear to be going toward warmer winter quarters. Rising at a considerable angle from the ground, on stiff, set, short wings, after a flushing, the birds, heading for a wooded cover, are off in a strung out line that only the tyro imagines makes an easy target. Suddenly dropping all at once and not far from each other, squatting close, in the confidence inspired by the perfect mimicry of their plumage with their surroundings, each bird must be almost trodden upon before it will rise to wing. Very rarely they take refuge in trees. It has been said a Bob White can retain its odor voluntarily, since the best of pointers often fails to find it even when within a few feet. When lying close, the wings are pressed against the side, every feather clings tightly with a tension produced by fear, in all probability, rather than by any voluntary act ; but the result is that by flying upward, rather than running and giving the scent to the dogs, and by compressing its feathers on dropping to the ground again, brave little Bob White often gives the sportsman a lively chase for his game. After much shooting, birds become "educated." Wonderfully clever they are in matching the sportsman's tricks with better ones. They school the wing shots finely until the crack marksman confesses his chagrin. The best trained dog may bushwhack an entire slope, where they are known to be scattered, without flushing one ; for vainly does the dog draw now. His usefulness was greatest in standing a covey before the reports from the gun gave fair warning that no one-sided sport had begun.

Once the firing ceases, sweet minor scatter calls—*quoi-hee, quoi-hee*—reunite the diminished members of a flock. A solitary survivor has been known to wander about the country through an entire winter, calling mournfully and almost incessantly for the missing brothers and sisters, until a farmer, whose family had feasted on their delicate white flesh, unable to listen to the cry that sounded to him like the voice of an accusing conscience, again picked up his gun and put the mourner out of misery.

Among the thousands upon thousands of "quail" shot annually, some sportsman finds either an albino or some other freak wearing plumage that he is certain belongs to a distinct species; but the Texan and the Florida birds alone are true, but merely climatic, variations of our own Bob White. The former is distinguished by its paler, more grayish tone of the upper parts, that are marked with tawny, while the Florida bird has darker, richer coloring, with heavier black markings, and a longer, jet black bill.

Several allied "quail" (partridges) are of too local a distribution on the Pacific slope and in the southwest to be included in a book that avowedly excludes "local and rare birds." Wherever the prolific Bob Whites have been introduced and protected in the west, they have so quickly spread as to encourage the hope that since true sportsmen everywhere are taking active measures to stay the hand of bird butchers, our national game bird may some day regain the vast numbers brutally destroyed.

Dusky Grouse
(Dendragapus obscurus)

Called also: BLUE, GRAY, MOUNTAIN, PINE, AND FOOL GROUSE; PINE HEN

Length—20 to 24 inches; length variable.

Male—Upper parts blackish brown, finely zigzagged with slatey gray mixed with lighter brown, and sometimes coarsely mottled with gray, especially on wings; forehead dull reddish brown; back of head blackish, the feathers tipped with rusty; sides of head black; shoulders streaked with white; long feathers on sides have white ends and shaft stripes; throat white, finely speckled with black; under parts bluish **gray or slate, varied** with white on flanks and underneath. Tail rounded, the twenty broad feathers blackish-brown, **marbled with gray,** and broadly banded across end with slate gray; legs covered to toes with pale brown feathers; a comb over eye; bill horn color.

Female—Smaller, lighter, more mottled, or blotched with blackish and tawny or buff, the feathers generally edged with white; slate gray under parts, and tail broadly banded with same; the flanks tipped with white and mottled with black and buff.

Range—Rocky and other mountain ranges in western United States.

Season—Permanent resident.

Two variations of the dusky grouse, known as Richardson's grouse and the sooty grouse—constantly confused in reports—make it somewhat difficult to define the exact habitat of this splendid game bird, so well known in one form or another by sportsmen throughout the western half of the United States, from New Mexico to Alaska and British Columbia. The habits of all three birds being practically the same, their plumage differing chiefly in degrees of duskiness, and their boundary lines constantly overlapping, it is small wonder the untrained observer confuses both their names and ranges. The Rocky Mountains, from central Montana and southeastern Idaho to New Mexico and Arizona, eastward to the Black Hills, South Dakota, and westward to East Humboldt Mountains, Nevada, is the range set down for the dusky grouse by the A. O. U., and a more westerly district, including California, for the sooty grouse; while Richardson's bird confines itself chiefly to the eastern slope of the Rockies. The latter is to be distinguished by its rather longer, square tail, with broader feathers, only slightly banded with gray, if at all, and its blacker throat. The sooty grouse, even darker still, and with a broad band on its tail, is minutely freckled with gray and rusty on its upper parts and very dark lead color below; the hen being particularly richly marked with rusty red and chestnut brown.

Taking the place in the western sportsman's heart of the ruffed grouse cherished in New England and the middle states, the dusky grouse, very like it in some habits and tastes, is a much larger bird, covered with a dense suit of feathers to resist the extreme cold of high altitudes, and weighing between three and four pounds. Next to the sage cock, this is the largest grouse in the United States. Possibly because it is so cumbrous, but more likely because its haunts are far removed from men, keeping it in ignorance, far from blissful, of his passion for hunting birds, this long-suffering recluse appears stupid to many. "Until almost fully grown," says a Colorado observer, "they are very foolish; flushed, they will tree at once, in the silly belief that they are out of danger, and will quietly suffer themselves to be pelted with clubs and stones until they are struck down one after another. With a shot gun, of course, the whole covey is bagged without much trouble; and as they are, in my opinion, the most delicious of all grouse for the table, they are gathered up unsparingly." When carnage like this masquerades under the title of "sport," evidently the extinction of the blue grouse, like that of many another choice game bird, is imminent. From an altitude of about seven thousand feet to timber line, coming down to the side hills and lower gulches, where food is more abundant for young broods in summer, the dusky grouse usually haunts rough slopes covered with dense forests of spruce and pine, and neither migrates nor strays far from its birthplace, though constantly roving. Solitary for part of the year, or found in small parties of three or four adults at most, it is chiefly while the young are partly dependent on the mother—for the male is an indifferent father—that one meets a covey of from seven to ten feeding on bearberries, raspberries, and other wild fruits, insects, especially grasshoppers, tender leaves, and leaf buds,

DUSKY GROUSE.
½ Life-size.

reserving the buds of the pine and the scales or seeds of its cones for winter fare, when nearly all other food is buried under snow. Heavy snowfalls send the grouse to roost in the evergreens, their dusky plumage, that blends perfectly with the sombre coloring of the pines as they squat on the limbs, making them all but invisible. Only early in the summer, when the young are unable to fly into the branches, do these tree-loving mountaineers roost on the ground. Approach a covey suddenly, and the beautiful, downy, nimble-footed chicks, that are by no means fools, scatter and hide among the bushes and under leaves, while the mother, flying in an opposite direction, alights in a tree, quite as if she had no family to be looked for; so why waste time in the search when she is in evidence? Moving her head from side to side, and looking at the disturber of her peace with first one eye, then the other, she will remain squatting on the limb just overhead with apparent apathy, or what passes for stupidity, but what may be the most intelligent self-sacrifice for her brood. Molest her, and she flies away very rapidly with a loud cackle of alarm. It is she that forms a depression in the ground, near an old log, in the underbrush, or in the stubble of an open field just as likely, but never far from water, after pressing down some fine grass, pine needles, or leaves to line the rude cradle. A clutch consists of from eight to ten creamy, buff eggs, dotted, spotted, and sometimes blotched with brown. Confining herself very closely for three weeks or longer, she at length leads forth a brood in June to call it by clucks and otherwise care for it precisely as the domestic hen looks after her chicks. The nesting begins about the middle of May, though dates differ with the severity of the season and the altitude. Only one brood is raised in a year.

While there is anything like work connected with raising a young family the father absents himself, to rejoin it only when the covey has agreeable society to offer and makes no demands. Yet this is the cock that in the mating season gave himself the airs of a turkey gobbler as he strutted along the mountain road in front of your wagon, tail spread to its fullest, wings dropped until they trailed over the ground—a picture of self-importance. This is the season when he woos his mate with booming thunder on a small scale, which passes for a love song. A small sac of loose, orange-colored skin, surrounded by a white frill of feathers edged with dusky, at either side of the neck, may now be inflated at will; and as the air escapes, a strange grumbling, groaning sound comes forth, seemingly from quite a distance, when perhaps very near, or, at least, from just the direction that it seems not to come from. This sound, that has been aptly likened to the distant laboring of a "small mountain sawmill wrestling in agony with some cross-grained log," may be uttered from a stump or rock, or in the air as the cock flies about from limb to limb of the evergreens. When disturbed, he has the habit of erecting the feathers on the back of his neck, a feeble showing as compared with the imposing black frill of the ruffed grouse.

Canada Grouse

(Dendragapus canadensis)

Called also: SPRUCE, WOOD, AND SPOTTED GROUSE; BLACK, SWAMP, AND SPRUCE PARTRIDGE; BLACK-SPOTTED HEATH COCK.

Length—14 to 15 inches.

Male—Upper parts ashy waved with black, gray, and grayish-brown. A few white streaks on shoulders; tail black, slightly rounded, and tipped with orange-brown; under parts black and white, the black throat divided from the black breast by a mottled black and white and ashy circular band; flanks pale brown, mottled or lined across with black; legs feathered to toes; bill black; a yellow or reddish comb over eye.

Female—Upper parts barred with black, gray, and buff, or pale rufous, the black predominating, except on grayish lower back; tail black, mottled and more narrowly tipped with orange brown than male's; under parts tawny barred with black; sides mottled with black and tawny; below black, the feathers broadly tipped with white.

Range—From northern New England, New York, Michigan, and Minnesota, westward to Alaska, and north so far as trees grow.

Season—Permanent resident; not a migrant, although a rover.

Only along the northern boundary of the United States may one hope to meet this small, hardy grouse walking about with the nimble steps of a Bob White, over the mossy bogs, in groves of evergreens and thickets of hackmatack—everywhere its favorite haunt; but in Canada it becomes increasingly abundant, and the *habitants* and *voyageurs* who penetrate the dark, swampy forests far to the north know it with that degree of intimacy which—perhaps because it furnishes the most interesting stories, that are at once the admiration and the despair of city-bred ornithologists—is discredited by them as "unscientific." There is a French Canadian, a native of the Laurentian Mountains, whose fleet ponies take many Americans to the Grand Discharge for the *oua naniche* fishing, who will lead his patrons to a nest beside a fallen log, show them the "drumming trees" where the cocks fly down and captivate their mates with a noise resembling distant thunder, point out a dusky figure in the sombre evergreens that no untrained eye could find as the buckboard rattles swiftly over the corduroy road, and at the camp-fire needs little persuasion to tell more about the Canada grouse than can be learned in the books.

Very early in spring the cocks begin to strut and give themselves grand airs. At this season especially, although the birds are never shy, the male exposes himself before an admiring observer with amusing abandon. With tail well up, and contracted and expanded at each step until the quills rustle like silk; with drooped wings, head erect, the black and white breast feathers standing out in regular rows, and those in the back of the neck correspondingly depressed; the combs over each eye enlarged at will and glowing red—a miniature impersonation of self-conceit struts through the forest, across one's path, flies into a low limb to attract more attention to his handsome body, and has been known to alight on a man's shoulder and thump his collar! Ordinarily he thumps any hard substance with his bill. Sometimes, with plumage arranged as above described, he will sit with his breast almost touching the earth and make peculiar nodding, circular motions of the head. To drum, he chooses some favorite tree inclined away from the perpendicular, and, commencing at the base, flutters slowly upward, very rapidly beating his wings to make the rumbling noise. Then, having ascended fifteen or twenty feet, he glides quietly to the ground, struts, and repeats the noisy ascent. A good "drumming tree," well known to woodsmen, often has its bark worn by the small thunderers. Apparently there are many more cocks than hens in every tamarack swamp.

Mr. Watson Bishop, of Nova Scotia, who succeeded in domesticating this grouse, tells many interesting fresh facts about it. A sitting hen, after scratching a depression in the ground, first lays three or four eggs before placing any nesting material in the cavity; then she has the absurd habit of picking up straws, leaves, etc., as she leaves the nest, and tossing them backward over her head, to land perhaps on the nest, or perhaps just in the opposite direction if she has faced about with head toward the eggs to secure some inviting material. When a quantity of litter has been collected, she will then sit on the eggs, reach out to gather it in and place it about her until the cradle is very deep and nicely bordered with grass and leaves. Jealousy, a ruling passion with hens at the nesting season, often leads them to steal one another's eggs. One nest should properly contain about a dozen, more or less, the ground color buff or pale brown, the spots and speckles reddish brown or umber; but so great is the variation of color and markings that some eggs have no markings at all, while others are beautifully and clearly decorated. It is possible to rub or wash off markings from many fresh-laid eggs. Laying commences about the first week of June; incubation lasts seventeen days, and by the middle of July the precocious chicks are able to reach the low branches of the evergreens in their first flights and move about on them like the adults that would make expert tight-rope walkers. Tender terminal spruce buds, hackmatack needles, the berries of Solomon's seal, pine needles and cones, and such fare give this grouse's flesh a dark color and a bitter, resinous flavor that tempts only the hungriest woodsmen; although in the berry season, when the birds leave the evergreens to feed on tender leaf buds and fruit, the rich reddish meat is much sought. An immense quantity of gravel is swallowed to aid digestion. Indians tell of following great packs of these grouse that furnished meat to a tribe for weeks; but a bevy of five or six birds is the largest recorded by scientists.

Ruffed Grouse

(Bonasa umbellus)

Called also: PARTRIDGE; PHEASANT; BIRCH PARTRIDGE

Length—16 to 18 inches.

Male and Female—Upper parts chestnut varied with grayish and yellowish brown, white, and black; head slightly crested;

RUFFED GROUSE

yellow line over eye ; sides of neck of male with large tufts of glossy greenish black feathers tipped with light brown, much restricted or wanting and dull in female ; long tail, which may be spread fan-like, yellowish brown or gray or rusty, beautifully and finely barred with irregular bands half buff, half **black** ; a broad subterminal band of black between gray **bands** ; throat and breast buff, the former unmarked ; underneath whitish, all barred with brown, strongly on **sides**, less distinctly on breast and below ; legs feathered to heel ; bill horn color.

Range—Eastern United States and southern Canada **west to** Minnesota, south to northern Georgia, Mississippi, **and Arkansas.**

Season—Permanent but roving resident.

Neither a "partridge" nor a "pheasant," it is by the former name that this superb game bird is best known to the New Englanders, and by the latter that it is commonly called in the middle and southern states; but this most typical grouse (whose Latin name describes two striking characteristics : *Bonasus*, a bison, referring to the bellowing bull-like noise produced by the male; and *umbellus*, to the umbrella-like tufts on his neck) appears in literature and the market stalls alike as a "partridge," a misnomer shared by the Bob White, which strictly belongs to a race of European birds of which we have no counterparts on this side of the Atlantic. What's in a name? That which we call a grouse by any other name doth taste as sweet.

Partial to hill country interspersed with cultivated meadows and dingles, or to mountains, rocky, inaccessible, thickly timbered, and well watered with bush-grown streams, it is only rarely, and then chiefly in autumn, that coveys leave high altitudes to feed along the edges of milder valleys and enter the swamps. The dainties preferred include crickets, grasshoppers, the larvæ of caterpillars, beechnuts, chestnuts, acorns of the chestnut oak and the white oak, strawberries, blueberries, raspberries, elderberries, wintergreen and partridge berries with their foliage, cranberries, the bright fruit of the black alder and dogwood, sumach berries (including the poisonous varieties, which do the grouse no injury), wild grapes, grain dropped in the stubble of harvested fields, the foliage of many plants, and the leaf buds of numerous shrubs and trees—a varied *menu* indeed, responsible alike for the bird's luscious, tender flesh and its roving disposition.

Bob Whites, Grouse, etc.

The "drumming" of a male ruffed grouse, its most famous characteristic, is surely as remarkable a bird call as is heard in all nature. A thumping, rolling tattoo, like the deep, muffled beating of a drum, sonorous, crepitating, ventriloqual, admirably written down by Mr. Ernest Seton Thompson, *thump————thump ————thump—thump, thump; thump, thump-rup, rup, rup, r-r-r-r-r-r-r-r*, announces the presence of a cock hopeful of attracting the attention of some shy female hidden in the underbrush. Any one will do, for he is a sadly erring mate, a flagrant polygamist, in spite of much that has been said to whiten his character. On a fallen log, a wall, or broad stump that has been used as a drumming ground perhaps for many years, and well known to the hens as a trysting place, the male puffs out his feathers until, like a turkey cock, he looks twice his natural size, ruffs his neck frills, raises his crest, spreads and elevates his tail, droops his trailing wings beside him, and, with head drawn backward, struts along the surface with the most affected jerking, dandified gait. Suddenly he halts, distends his head and neck, and beats the air with his wings, slowly at first, then faster and faster, until there is simply a blur where wings should be, so marvelously fast do they go. Because they vibrate at a speed at which the human eye can scarcely follow, the method of drumming is a vexed question among the most reliable observers. Thoreau was ready to swear that he had seen the ruffed grouse strike its wings together behind its back to produce the sound, Audubon to the contrary notwithstanding. Most woodsmen will tell you either that the male strikes the log on which he is standing, or the sides of his body; but the strongest scientific judgment now favors the abundant testimony that the bird beats nothing but the air; its wings neither meet behind the back, nor do they touch its sides, nor strike against any substance whatsoever. The drumming may occur at any season, most frequently and vigorously at nesting time, of course; but besides being a love "song," it is doubtless also a challenge to rival cocks, that fight like gamesters until blood and feathers strew the ground; or it may be simply an outlet to the bird's inordinate vanity and vigorous animal spirits. In a lesser degree the sound is precisely the same as when the grouse begins its flight.

Quite ignored by her lover when maternal duties approach, the female scratches a slight hollow in some secluded place, usually at the foot of an old stump or log or rock, often near a stream among the underbrush; but many nests in unprotected open stretches are recorded. A few wisps of dry grass, dead leaves, pine needles, or any convenient material, line the hollow in which a full set of eggs—from ten to fifteen rich buff, dotted with different sized spots of pale chestnut brown—has been found as early as April first, a full month earlier than the regular time. Since the markings can be easily rubbed off a fresh laid egg, one sometimes hears that the grouse's egg is plain buff. Only one brood is raised in a season, the exceptions to the rule being very rare. For nearly four weeks the hen closely confines herself, and, like the sitting Bob White, relies upon her plumage's perfect mimicry of her surroundings to protect her from notice. The coloring of a ruffed grouse tells of a long ancestry passed under deciduous trees. Seated among last year's leaves she looks all of a piece with the carpeting of the woods, and neither stirs a feather nor winks an eye, though you stand within two feet of her, to lead you to think otherwise. Mr. D. G. Elliot, among others, believes she hides her nest from the male as well as from all her other enemies. The fox, weasel, squirrel, hawk, owl, and above all the breech-loader, are the grouse's deadliest foes; and a species of woodtick that inserts its triangular head beneath the skin, sometimes destroying entire broods. Bird lice, and a botworm that resembles a maggot and penetrates the flesh, likewise prove fatal, particularly to chicks. The dust baths commonly indulged in are taken to rid themselves of vermin. Heavy rains that drench the fledgelings not infrequently kill them, too, until one wonders there are any ruffed grouse left. The precocious, downy brown balls, that run at once from the shell, are managed precisely as a domestic hen cares for her brood, even to the clucking, hen-like call that summons them beneath her wings, where they sleep until old enough to roost in trees like adults. The mother grouse when suddenly startled gives a shrill squeal, apparently the signal for the covey to scatter and hide among the leaves and tangle, while, by feigning lameness and other hackneyed devices for diverting an intruder's attention from the chicks to herself, she remains in their neighborhood, they motionless in their hiding places, until the reassuring *cluck* calls the happy family together again. When the young need no further care in autumn, the males selfishly join the covey, rarely consisting of more than six or eight birds; for, unlike the pinnated grouse, this species does not pack.

"The ruffed grouse, by reason of its sudden bursts from cover, its bold, strong, swift flight, the rugged nature of its favorite cover, its proud, erect carriage, its handsome garb and its wide distribution is easily the king of American game birds," says Mr. G. O. Shields, "and has therefore been chosen as the emblem of the League of American Sportsmen."

In the brisk, golden days of autumn the sportsman finds sport indeed in hunting the wily, clever grouse, "educated" by much persecution from an almost tame denizen of the mountain farm into a woodland recluse that constantly challenges admiration for its cunning. It will seldom lie well to a dog, but sneaks away so swiftly through the underbrush that either the dog or its master usually gets left. By flying low, then dropping to run again, the strong scent is broken.

Bob Whites, that have a power of withholding their scent by tightly compressing their feathers—a trick not known to the grouse apparently—do not escape detection any better than they. Many skilled sportsmen, armed with the most approved breech-loaders, and aided by the best trained dogs that bushwhack a region where grouse are known to be abundant, return home with light bags. No bird that flies, unless it is the Jack snipe, is so seldom hit. A tremendous *whir-r-r-r* of rapidly beaten wings startles the tyro out of a good aim. Unusually strong chest muscles for concentrated but limited exertion, and especially stiff wings, enable the grouse to hurl themselves into the air with a thunderous velocity; but, like all their allies, they can steal away as silently as Arabs, if necessary. Darting away directly opposite from the sportsman, a well "educated" bird quickly places a tree between itself and the shooter, threading a tortuous maze in and out through the woods, higher and higher, until, having cleared the tree tops, it is off to freedom. Fear, not a natural, but an acquired state of mind, has not yet blasted the peace of grouse in regions where they have never been molested; and knowing no worse enemy there than a fox, from which they are safe when roosting in a tree, and mistaking the sportsman's dog for one, they have been sometimes credited with stupidity because by remaining on the perch they allow a man to rake the covey. But such assault and battery is happily rare. Certain hawks and owls do awful execution. Snares of silk and horsehair, poacher's traps, and "twitch-ups" of young saplings bent by the farmer's boy, do much to spoil the sport, that becomes shockingly rarer year by year. To escape pursuit a grouse will often dive into the snow; and although dense feathers cover its body and legs, it will make a similar plunge to keep warm in extremely cold weather, a solitary shiverer, unlike the Bob Whites, that bury themselves in cosy, snug family parties; but, like them, it, too, sometimes gets imprisoned by an impenetrable ice crust, and so perishes miserably.

. . . .

The Canadian Ruffed Grouse (*Bonasa umbellus togata*), to be distinguished from the preceding by the prevailing gray, instead of chestnut, of its upper parts, its grayer tail, and its more distinctly barred under parts, almost as clear on the breast and underneath as on the sides, is doubtless simply a climatic variation, only the systematists seeing a sufficient difference in the two birds to justify their separation into two distinct species. Their habits and eggs are identical. Often no difference can be detected by sportsmen who bring home both species in their game bags. The spruce forests of northern New York and New England and the British provinces, westward to Northern Oregon, Idaho, and Washington to British Columbia, north to James Bay, is the Canadian ruffed grouse's range.

. . . .

The Gray Ruffed Grouse (*Bonasa umbellus umbelloides*), a still paler variation, in which the gray tints predominate, ranges from the Rocky Mountain region of the United States and British America north to Alaska and east to Manitoba. Considering the altitudes of from seven to ten thousand feet at which it usually lives, the lonely cañons it frequents, and its rare persecution at the hands of men, it is surprisingly shy, according to Captain Bendire. Otherwise it has no trait, apparently, not already touched upon in the life history of the ruffed grouse.

. . . .

The Oregon, or Red Ruffed Grouse (*Bonasa umbellus sabini*), the darkest, handsomest variation of the ruffed grouse anywhere found, roams over the coast mountains of Northern California, Oregon, Washington, and British Columbia, reaching Alaska and many of the Pacific Coast islands, and occasionally straying into

Colorado, Dakota, Montana, and Idaho. Where the Canadian variety encroaches its territory, however, little or no difference in the plumage may be detected. The account of the ruffed grouse's habits, nest, etc., should be read to avoid repetition, since the Oregon bird is simply a climatic variation of the eastern species.

Prairie Chicken
(Tympanuchus americanus)

Called also: PINNATED GROUSE; PRAIRIE HEN

Length—About 16 to 18 inches.

Male and Female—Upper parts brown, barred with black, chestnut, ochre, and whitish, the latter chiefly on wings; sides of the neck tufted with ten or more narrow, stiff feathers, rounded at end, which may be erected like conventional Cupid's wings above the head. Their color black, with buff centres, frequently chestnut on inner webs; bare, yellow, loose skin below these feathers may be inflated at will; the dusky, brown, white tipped tail rounded, the inner feathers somewhat mottled with buff; chin and throat buff; breast and underneath whitish, evenly barred with black. Head slightly crested; legs scantily feathered in front only. Female smaller, the neck tufts much restricted, no inflated sacs below them; the tail feathers with numerous distinct buff bars.

Range—"Prairies of the Mississippi Valley; south to Louisiana and Texas; east to Kentucky, Indiana, Ohio, Michigan and Ontario; west through eastern portions of North Dakota, South Dakota, Nebraska, Kansas, and the Indian Territory; north to Manitoba; general tendency to extension of range westward and contraction eastward; migration north and south in Minnesota, Iowa and Missouri."—A. O. U.

Season—Permanent resident; only locally a migrant at northern limit of range.

Westward the prairie chicken, like the course of empire, takes its way; for although it may increase at the pioneer stage of civilization, it halts at the introduction of the steam plough and railroad, to disappear forever where villages run together into cities. Doubtless its range was once far east—just how far is not certain, since the early writers confused it with the heath hen, once

PRAIRIE HEN.

enormously abundant, but now confined to Martha's Vineyard, where in 1890 there were about one hundred of the birds left, and now, for the want of sufficient protection, even this pitiful remnant has diminished to very near the extinction point. So it will be inevitably with the prairie chicken. Modern farming machines destroy thousands of eggs and young annually as they steam over the prairies; in the small, new settlements there is little respect paid to game laws when a dull monotony of salt pork sets up a craving for fresh meat; and since the prairie chicken has strong preferences for certain habitats, and will not or cannot live in others, evidently the day is not far distant when either missionary effort on behalf of this and many other birds must be vigorously applied, or they will certainly perish from the face of the earth. Since the coyote, or prairie wolf, which has preyed on this grouse, is being killed off, and sportsmen are endeavoring to enforce the law against trapping the birds in winter, and to induce farmers to burn off their fields in autumn instead of in May, there is still hope that its extinction may be at least postponed.

Early in the morning in spring the booming of males assembled on the "scratching ground"—some slight elevation of the prairie—summons the hens from that territory to witness their extraordinary performances until the whole region reëchoes with the soft though powerful sound, like deep tones from a church organ—harmonious, penetrating, more impressive to the human listener than to the apparently indifferent females. Inflating the loose yellow sacs on the sides of their head, that stand out like two oranges; erecting and throwing forward their Cupid-like feathers at the back of the neck; ruffling the plumage until it stands out straight; drooping the wings and spreading the erect tails, the males present an imposing picture of pompous display and magnificence that melts not the flinty hearts of the coquetting spectators. Now the proud cock, incited to nobler deeds by the indifference of his chosen sweetheart, rushes madly forward, letting the air out of his cheek sacs as he goes, to produce the booming noise, repeating the rush toward her and the *boom* until she gives some sign that his mad endeavors to win her awaken some response in her cold little heart. Toward the end of courtship she moves about quickly among the performers, then stands perfectly still for a time, evidently taking note of the fine points of the numerous lovers that embarrass her choice. Shortly after the sun rises, the circus and concert end for the day, to be repeated the next morning, and the next, for a week or longer, at the end of which time the inflamed cocks usually fall to fighting, clawing at each other as they leap into the air and scatter blood and feathers. To the victor belongs the sweetheart. The note of the male bird is closely imitated by many farmers' boys. It may be written, *uch-ah-umb-boo-oo-oo-oo*.

It must be owned these birds show no great intelligence in the selection of nesting sites, large numbers of homes placed in the short grass of dry localities being destroyed by prairie fires annually, others on cultivated lands are crushed by mowing machines, and those built along the marshes or sloughs are often inundated in a wet season. A slight excavation, sometimes thickly, but more often sparsely, lined with grasses and feathers plucked from the mother's body, receives from ten to twenty eggs, ranging from cream to pale brown, regularly marked with fine reddish brown dots, the coloring and spotting differing, however, on almost every egg in a clutch. It is the female that bears the entire burden of incubation, lasting from twenty-three days to four weeks. So perfectly does her plumage mimic her surroundings that one may almost step on a nest without seeing her. Like all her tribe, she is a model mother, she alone caring for the downy chicks, leading them where grasshoppers and other insect fare abounds, and protecting them with courageous and artful tactics.

The young are marvelously cunning in hiding in the grass. Now they lie very close to a dog, and since their flesh is white and toothsome, whereas that of old birds is dark and less esteemed, they fill the game bags after the fifteenth of August. Toward the end of summer, when there is no nursery work left to do, the selfish father joins his family; other families join his, or pack, until in regions where the birds have not been persecuted several scores roam over the prairie together to feed in the grain fields and on small berries and seeds. Now the grouse become wilder, and, except when gorged to indolence, will fly a mile or more, perhaps, so that little sport can be had with them over dogs.

"The true manner of shooting prairie fowl," says Mr. Charles E. Whitehead in "Sport with Rod and Gun," "is to drive over the prairie in a light wagon, letting the dogs range far and wide on either side. . . . When one scents the birds he will come to a point suddenly . . . as if he saw a ghost. The wagon drives near him, the other dogs coming up and backing him. The sportsmen then alight and take their shots. Rarely the whole covey is flushed together, and frequently the old birds lie until the last, and while the sportsman is loading his gun will dash away uttering their quick-repeated cry of *cluk-cluk-cluk-cluk*, and looking back over their wings at the sportsman who marks them down half a mile away. As one goes to retrieve the dead bird, still another and another will rise, and it is only until one has been carefully over the field that he feels secure that all the birds are up."

Unlike the rest of their kin, the prairie chickens can fly long distances, though not with such concentrated power as to produce the thunder-like roar of the ruffed grouse, for example. Their flight may not be so swift, for it is accomplished with less flapping and more easy, graceful sailing. They migrate regularly, or, at least, the females do, leaving the hardier males to brave the intense cold at the northern limit of their range. In November and December flocks descend from northern Iowa and Minnesota to settle for the winter in southern Iowa and northern Missouri, the size of the south bound flocks being influenced by the severity of the cold, just as the return of the migrants in March and April depends upon the warmth of spring. Most of the pinnated grouse's life is passed on the fertile open prairies, sleety storms, high winds, and deep snow alone driving a pack to shelter in timbered lands.

Prairie Sharp-tailed Grouse

(Pediocætes phasianellus campestris)

Called also: PIN-TAILED GROUSE ; SPECKLED OR WHITE BELLY; WILLOW GROUSE ; PRAIRIE CHICKEN; SPIKE-TAIL

Length—17.50 to 20 inches.

Male and Female—Upper parts yellowish buff, irregularly barred and blotched with black; the shoulders streaked and the tips of wing coverts conspicuously spotted with white; crown and back of neck more finely barred than the back; no neck tufts; head of male slightly crested, and his neck has concealed reddish distensible skin; space in front of and below eye buff, like the throat; breast has V-shaped brownish marks; sides irregularly barred or spotted with blackish or buff; underneath, including wing linings, white. Tail barred with black and buff, the central feathers longest, but shorter in female than in male; legs full feathered to the first joint of toes; bill horn color. Female smaller.

Range—Plains and prairies east of the Rocky Mountains, north to Manitoba, east to Wisconsin and Illinois, and south to New Mexico.

Season—Permanent resident, or partially migratory in cold weather.

Three variations of one species of sharp-tailed grouse greatly extend its range until in one form or another it has come to be among the best known of our western game birds; the Columbian, the true sharp-tail, and the prairie varieties not being generally separated by sportsmen either in the United States or Canada, as they are by the systematists.

A most hilarious "dance" that precedes the nesting season, as in the case of the pinnated grouse, begins early in spring, at the gray of dawn, when the sharp-tails meet on a hillock that very likely has been a favorite with their ancestors, too. They behave like rational fowls until suddenly a male lowers his head, distends the sacs on either side of his neck that look like oranges fastened there, ruffles up his feathers to appear twice his natural size, erects and spreads his tail, droops his wings, and, rushing across the arena, "takes the floor." Now the ball is opened indeed. Out rush other dancers, stamping the ground hard as their feet beat a quick tattoo, the air escaping from their bright sacs making a "sort of bubbling crow," quite different from the deep organ tone of the pinnated grouse; the rustling of the vibrating wings and tail furnishing extra music. Now all join in; at first there is dignified decorum, but the fun grows fast and furious, then still faster and still more furious; the crazy birds twist and twirl, stamp and leap over each other in their frenzy, every moment making more noise, until their energy finally spent, they calm down into sane creatures again. They move quietly about over the well worn space (a "chicken's stamping ground," measuring from fifty to one hundred feet

PRAIRIE SHARP-TAILED GROUSE.
⅓ Life-size.

across, according to Mr. Ernest Seton Thompson), when, without warning, some male has a fresh seizure that soon starts another saturnalia. "The whole performance reminds one so strongly of a 'Cree dance,'" says Mr. Thompson, "as to suggest the possibility of its being the prototype of the Indian exercise. . . . The dancing is indulged in at any time of the morning or evening in May, but it is usually at its height before sunrise. Its erotic character can hardly be questioned, but I cannot fix its place or value in the nuptial ceremonies. The fact that I have several times noticed the birds join for a brief 'set to' in the late fall merely emphasizes its parallelism to the drumming and strutting of the ruffed grouse as well as the singing of small birds."

After pairing, the male, in the usual selfish fashion of his tribe, allows his mate to seek some place of concealment, scratch out an excavation screened by grasses, and attend to all nursery duties, while he joins a club of loafers that most scientists consider flagrant polygamists too. From ten to sixteen eggs, very small for so large a bird, and of a brown or buff shade with a few dark spots, hatch, after about twenty-one days of close sitting, into golden yellow, speckled chicks, admirably clothed, to escape detection from prowling hawks, as they squat in the grass. This species, too, is a conspicuous sufferer from the mowing machine and prairie fire. If farmers would only burn all their fields in autumn instead of in May and June, when birds are nesting, thousands of grouse might be spared annually.

All young grouse feed largely on insects, especially grasshoppers, at first, but sharp-tails become almost dependent at any time on the hips of the wild rose, the stony seeds that likewise do the work of gravel being a staple every month in the year; willow and birch browse, various seeds, cereals, and berries enlarging a long *menu*. Such dainty fare makes delicate, luscious flesh, so tender, indeed, that young birds falling at the aim of the sportsman's gun have been burst asunder when they reached the ground, and their feathers loosened. With increased age the flesh grows dark and less palatable. These grouse, hunted in the same fashion that the pinnated grouse is, generally lie well to a dog. A single bird rising with a cackling cry when flushed at a point, flies swiftly straight away, now beating the wings, now sailing with them stiffly set and decurved, still cackling as it goes. Later in the year, when coveys unite to form a dense pack, the eyes, turned in all directions at once, on the perpetual lookout, it is a skilled sportsman who can steal a march on them before they run swiftly away and finally take to wing to flap and sail far, far beyond reach of his gun. When cold blasts, high winds, and deep snow drive these prairie lovers into timbered lands and sheltered ravines, a covey spends much time roosting in trees and walking along the branches, where the sharp-tails' nature apparently undergoes a change; for it is said they are almost stupidly unsuspicious now, and will sit still and look on at the destruction of their companions. Odd that they should shun man and his habitations! A partial migration of females to warmer, or at least more sheltered winter quarters, doubtless accounts for the variation of the species.

The Columbian Sharp-tailed Grouse (*Pediocætes phasianellus campestris*), also called by the various popular names by which the prairie sharp-tail is known since few see any difference between the two varieties, has its upper parts more grayish instead of yellowish buff, possibly with less conspicuous white spots on its wings and shoulders, and its whitish under parts, including flanks, marked with black U or V shaped lines. In habits there appears to be little or no difference between this variety and its prototypes; therefore the account of the prairie sharp-tail need not be repeated. As its name implies, the region about the Columbia River is this grouse's chosen habitat; but the northwestern part of the United States, including northeastern California, northern Nevada, and Utah, Montana, Wyoming, Oregon, Washington, and from west of the Rocky Mountains northward through British Columbia to central Alaska, is the area over which it is distributed. As man, whom it shuns (unlike the pinnated grouse), appears on its territory, it recedes before him into wilder, remote districts, until plains where coveys were abundant only five years ago now know them no more.

The Sharp-tailed Grouse (*Pediocætes phasianellus*), a bird that never shows its dark, rich plumage within the United States, however commonly the paler, yellower prairie, and the grayer Columbian varieties of this handsome grouse are called by its name, ranges over the interior of British America to Fort Simpson, and is comparatively little known. Reversing the usual rule, the plumage of this one species grows gradually darker as the birds range northward, until the true sharp-tail has black for its prevailing color.

Sage Grouse

(Centrocercus urophasianus)

Called also: SAGE COCK; COCK OF THE PLAINS

Length—20 to 32 inches; largest of the grouse.

Male—Upper parts ashy gray barred with brown, black, and darker gray; some white streaks on wings; tail of twenty stiff feathers graduated to a threadlike point, the central ones like back, the outer ones black and partly barred with buff; top of head and neck grayish buff. ("Neck susceptible of enormous distention by means of air sacs covered with naked, livid skin, not regularly hemispherical and lateral like those of the pinnated grouse, but forming a great protuberance in front of irregular contour; surmounted by a fringe of hairlike filaments several inches long, springing from a mass of erect, white feathers; covered below with a solid set of sharp, white, horny feathers like fish scales. The affair . . . is constantly changing with the wear of the feathers."—Dr. Elliott Coues). This neck decoration is fully displayed only at the pairing season. Fore neck black speckled with grayish; breast gray; flanks broadly barred with blackish brown and pale buff, or sometimes mottled with black; underneath black; wing linings white.

Female—One-third smaller than male; chin and throat white; no neck decoration; a softer, shorter tail.

Range—Sage covered and sterile plains of British Columbia, Assiniboia, the two Dakotas, Nebraska, Colorado, southward to New Mexico, Utah, and Nevada; west to California, Oregon, and Washington.

Season—Permanent resident, or partly migratory at some points.

Several peculiarities make this species noteworthy; next to the turkey it is the largest game bird in the United States, as it is the largest of the grouse clan, a full grown male weighing often eight pounds, while his smaller mate may be only a little over half that weight, the size of sage fowls differing greatly. Another distinction it possesses in being the only one of the gallinaceous or scratching birds without a gizzard, what answers for one being merely a soft, membraneous bag; hence gravel, prairie rose seeds, and other hard substances are never swallowed. Because sage grouse are commonly found in regions where the bush that lends them its name abounds, there is a popular impression that its leaves are their sole diet; but while they certainly form its staple in winter, at least, immense numbers of grasshoppers, crickets, berries, grain, seeds of grasses, and leguminous plants so change the character of the bird's flesh, ordinarily bitter and astringent, as to make it truly palatable to the fastidious in many sections where only the sharpest appetite could relish it under the sage circumstances. But even then a young bird should be drawn immediately after death.

Since the sage bush *(Artemisia)* grows to a height of only two or three feet, a partial migration of a winter pack sometimes becomes necessary when the plant is hopelessly buried under snow, however willing this as well as other grouse may be to plunge into shallow drifts. Intense cold, common to the high altitudes, and intense heat to the alkali regions it inhabits, blizzards or scorching winds, apparently do not affect this hardy bird. The food supply is its first consideration; after that a drink morning and evening from some clear mountain stream. At the approach of winter, coveys of seven or eight birds begin to pack into flocks, sometimes numbering a hundred, whose strong, clannish feeling leads them to live much as the Bob Whites do, though the males are no such models of the domestic virtues. Forming in a circle, the grouse squat and huddle for mutual warmth and protection, tails toward the centre of the ring, heads pointing outward to detect danger that may come from any direction. Yet they are not suspicious birds, or wild; they generally walk quietly away from an intruder, or run and hide among the sage bushes, where, owing to the mimicry of their plumage, it is difficult indeed to detect them. Their nature is terrestrial. Flying, at the outset a laborious performance, will not be resorted to except as a last expedient. The sage cock with effort lifts his heavy body from the ground by much wing flapping; his balance is unsteady until fairly launched; but once off, on he goes, alternately flapping with five or six quick strokes, then smoothly sailing, cackling his alarm as he flies, until far beyond sight. Wheat found in the crop of a bird killed early in the morning eight miles from a cultivated field, proves to what a distance this grouse is willing to fly for a good breakfast. Mr. D. G. Elliot says it requires a heavy blow to bring a bird down, large shot being necessary to kill one; for it is capable, even if severely wounded, of carrying away large quantities of lead, and will fly a long distance, probably not dropping until life is extinct. Like the prairie hen and the sharp-tailed grouse, only one bird will flush at a time, the others lying close in concealment.

Like these birds, too, the sage cock goes through some amusing pre-nuptial performances early in spring. Inflating his large saffron colored air sacs until they rise above his head and all but conceal it, the spring feathers along the edges standing straight out, his pheasant-shaped tail spread like a great, pointed fan, the wings trailing beside him, his breast rubbing the ground until often the feathers are worn threadbare, he moves around the object of his affections with mincing, gingerly steps, while the air escaping from the sacs produces a guttural, purring sound that seems to voice his entire satisfaction with himself. Notwithstanding his protestations of devotion, he leaves his mate to scratch out a nest under some sage bush or in a grass tussock, and here she confines herself very closely—for she is a model mother—for three weeks or more. Knowing how perfectly her feathers conceal her from the sharpest eyes, she remains on the nest until sometimes almost stepped on, and shows the marvelously clever tricks of protecting her chicks common to all this highly intelligent clan. It is the coyote that is her deadliest enemy. When the brood is fully able to take care of itself, the neglectful father, that has passed the early summer with other cocks as selfishly indolent as he, for the first time becomes acquainted with his children.

PHEASANTS AND TURKEYS

(Family Phasianidae)

Wild Turkey
(Meleagris gallopavo)

Length—About four feet; largest of the game birds.
Male—Head and upper neck naked; plumage with metallic bronze, copper, and green reflections, the feathers tipped with black; secondaries green barred with whitish, the primaries black barred with white. (The wild turkey to be distinguished from the domestic bird chiefly by the chestnut, instead of white, tips to the tail and upper tail coverts.) A long bunch of bristles hangs from centre of breast; bill red, like the head; legs red and spurred.
Female—Smaller, dull of plumage, and without the breast bristles.
Range—United States, from the Chesapeake to the Gulf coast, and westward to the Plains.
Season—Permanent resident.

Once abundant so far north as Maine, Ontario, and Dakota, this noble game bird, now hunted to very near the extinction point, has had its range so restricted by the advance of civilization, for which it has a well grounded antipathy, that the most inaccessible mountains or swampy bottom lands, the borders of woodland streams that have never echoed to the whistle of a steamboat, are not too remote a habitation. Originally no more suspicious and wild than a heath hen, according to the testimony of early New Englanders, much persecution has finally made it the most cunning and wary, the most unapproachable bird to be found; but what possible chance of escape has any wild creature once man, with the manifold aids of civilization at his disposal, determines to possess it? It cannot be long at the present rate of shrinkage before the turkey, in spite of its marvelous cleverness, will follow the great auk to extinction.

WILD TURKEY.
¼ Life-size.

It is the Mexican turkey, introduced into Europe early in the sixteenth century, that still abundantly flourishes in poultry yards everywhere, and furnishes our Thanksgiving feasts. Another bird of the southwest, the Rio Grande turkey, that ranges over northeastern Mexico and southeastern Texas, and a fourth and smaller variety, confined to southern Florida, show constant, if slight variations in plumage, but little in nature, which awakens the hope that if American sportsmen were to introduce the southern races where the present species has been killed off, and protect the birds, magnificent sport might be indefinitely preserved.

Beginning at early dawn in spring, and before leaving his perch, the male turkey gobbles a shrill, clear love song, quite different at this season, before fat chokes his utterance, from the coarse gobble of the domestic turkey. The females now roost apart, but in the same vicinity. By imitating the hoot of the barred owl, and by skilful counterfeits of the female's plaintive yelp, produced by old sportsmen with the aid of a turkey wing-bone, or a vibrating leaf placed on the lips, among other devices, the turkey may be lured within gun range, if his education has not gone far. Sailing to the ground from his perch, in the hope of having attracted some hen to his breakfast ground, the cock, at sight of one, displays every charm he possesses : his widely spread tail, his dewlap and warty neck charged with bright red blood ; and drooping his wings as he struts before her, he sucks air into his windbag, only to discharge it with a pulmonic puff, that he evidently considers irresistibly fascinating. Dandified, overwhelmingly conceited, ruffled up with self-importance, he struts and puffs, until suddenly an infuriated rival rushing at him gives battle at once; spurs, claws, beaks, make blood and feathers fly, and the vanquished sultan retires discomfited, leaving the foe in possession of the harem. The turkey is ever a sad polygamist. Once the nesting season, lasting about three months, is over, the male stops gobbling, and not until the young need no care does he rejoin the females and see his well grown offspring for the first time, having enjoyed an idle club life with other selfish males while there was any real work to do.

The turkey-hen, happy in his exile, even takes pains to hide herself and nest from his lordship, for he becomes frightfully jealous of anything that distracts her attention from him, and will destroy eggs or chicks in a fit of passion. Evidently jealousy is unknown to her, however, for many nests—or the area of ground that answers as such—have been reported where two hens deposited their cream colored eggs, finely and evenly speckled with brown, thus doubling the ordinary clutch into one of two dozen eggs or over. It is thought that, in such cases, the good-natured incubators relieve each other. Snakes, hawks, and other enemies in search of so toothsome a morsel as a turkey chick, and heavy rains that chill the delicate, downy fledgelings, decimate a brood, however faithfully tended by a devoted mother. It is not until they are able to fly into high roosts that her mind is relieved of many anxieties ; and only when some dire calamity sweeps away her entire family does she attempt to raise a second brood. Insects, especially grasshoppers, appear to be the approved diet for all young gallinaceous fowl ; the more extensive bill of fare of fruits, grain, nuts, seeds, and leaf buds comes later, when a toughened gizzard may receive the quantities of gravel necessary to grind the grain. *Quit, quit*, call the feeding birds, though, like domestic fowls, to quit is the last thing they seem ready to do. Where food is abundant they may wander far, but never from a chosen region, for they are not migratory; nevertheless the pointer that scents a small flock in autumn, when the innocence of young birds makes shooting a possibility to the expert, leads his master a rough and wearisome chase before a shot is offered at this peerless game bird.

COLUMBINE BIRDS

Pigeons
Doves

COLUMBINE BIRDS
(Order Columbæ)

Pigeons and Doves
(Family Columbidæ)

Of three hundred birds of this order known to scientists, over one hundred are confined to the Malay Archipelago, yet there are only twenty-eight in India, fewer still in Australia, only twelve species in the whole of North America, and of that number but two pigeons and one dove now stray far enough **beyond** the Florida Keys and southern borders to come within the scope of this book.

> Passenger, or Wild Pigeon.
> Band-tailed Pigeon.
> Mourning Dove.

PIGEONS AND DOVES
(Family Columbidæ)

Passenger Pigeon
(Ectopistes migratorius)

Called also : WILD PIGEON

Length—16 to 25 inches.

Male—Upper parts bluish slate shaded with olive gray on back and shoulders, and with metallic violet, gold, and greenish reflections on back and sides of head; the wing coverts with velvety black spots; throat bluish slate, quickly shading into a rich reddish buff on breast, and paling into white underneath; two middle tail feathers blackish; others fading from pearl to white. Eyes red, like the feet; bill black.

Female—Similar, but upper parts washed with more olive brown; less iridescence; breast pale grayish brown fading to white underneath.

Range—Eastern North America, nesting chiefly north of or along the northern borders of United States as far west as the Dakotas and Manitoba, and north to Hudson Bay.

Season—Chiefly a transient visitor in the United States of late years.

The wild pigeon barely survives to refute the adage, "In union there is strength." No birds have shown greater gregariousness, the flocks once numbering not hundreds nor thousands but millions of birds; Wilson in 1808 mentioning a flock seen by him near Frankfort, Kentucky, which he conservatively estimated at over two billion, and Audubon told of flights so dense that they darkened the sky, and streamed across it like mighty rivers. So late as our Centennial year one nesting ground in Michigan extended over an area twenty-eight miles in length by three or four in width. The modern mind, accustomed to deal only with pitiful remnants of feathered races, can scarcely grasp the vast numbers that once made our land the sportsman's paradise. Union for once has been fatal. Unlimited netting, even during the entire nesting season, has resulted in sending over one million pigeons to market from a single roost in one year, leaving perhaps as many more wounded birds and starving, helpless, naked squabs behind, until the poultry stalls became so glutted with pigeons that the low price per barrel scarcely paid for their transportation, and they were fed to the hogs. This abominable practice of netting pigeons, discontinued only because there are no flocks left to capture, has driven the birds either to nest north of the United States, or, when within its borders, to change their habits and live in couples chiefly. Captain Bendire, than whom no writer ever expressed an opinion out of fuller knowledge, said in 1892: "The extermination of the passenger pigeon has progressed so rapidly during the past twenty years that it looks now as if their (sic) total extermination might be accomplished within the present century." Already they are scarce as the great auk in the Atlantic states.

One, or at most two white eggs, laid on a rickety platform of sticks in a tree, where they are visible from below, would scarcely account for the myriads of pigeons once seen, were not frequent nestings common throughout the summer; and it is said the birds lay again on their return south. Both of the devoted mates take regular turns at incubating, the female between two o'clock in the afternoon and nine or ten the next morning, daily, leaving the male only four or five hours sitting, according to Mr. William Brewster. "The males feed twice each day," he says, "namely, from daylight to about eight A.M., and again late in the afternoon. The females feed only in the forenoon. The change is made with great regularity as to time, all the males being on the nest by ten o'clock A.M. . . . The sitting bird does not leave the nest until the bill of its incoming mate nearly touches its tail, the former slipping off as the latter takes its place. . . . Five weeks are consumed by a single nesting. . . . Usually the male pushes the young off the nest by force. The latter struggles and squeals precisely like a tame squab, but is finally crowded out along the branch, and after further feeble resistance flutters down to the ground. Three or four days elapse before it is able to fly well. Upon leaving the nest it is often fatter and heavier than the old birds; but it quickly becomes thinner and lighter, despite the enormous quantity of food it consumes." Before it leaves the nest it is nourished with food brought up from the parents' crops, where, mixed with a peculiar whitish fluid, it passes among the credulous as "pigeon's milk." Is not this the nearest approach among birds to the mammals' method of feeding their young?

Patterns of all the domestic virtues, proverbially loving, gentle birds, anatomists tell us their blandness is due not to the cultivation of their moral nature, but to the absence of the gall-bladder!

. . . .

The Band-tailed, or White-collared Pigeon *(Columba fasciata)*, a large, stout species distributed over the western United States and from British Columbia to Mexico, inhabits chiefly those mountainous regions where acorns, its favorite food, can be secured. The male has head, neck, and under parts purplish wine red, fading below; a distinct white half collar, with some exquisite metallic scales on it; his lower back, sides of body, and wing linings slaty blue; the back and shoulders lustrous dark greenish brown; yellow feet and bill; a red ring around eye; and the bluish ash tail crossed at the middle with a black bar. The female either lacks the white collar or it is obscure, and her general coloring is much duller. Like the passenger pigeon, this bird sometimes lives in flocks of vast extent, its habits generally according with those previously described.

Mourning Dove[*]

(Zenaidura macroura)

Called also : CAROLINA DOVE ; TURTLE DOVE

Length—12 to 13 inches.

Male—Grayish brown or fawn color above, varying to bluish gray. Crown and upper part of head greenish blue, with green and golden metallic reflections on sides of neck. A black spot under each ear Forehead and breast reddish buff; lighter underneath. (General impression of color, bluish fawn.) Bill black, with tumid, fleshy covering; feet

[*] The account of the Mourning Dove, which, in a work scientifically classified, belongs in its present position, is reprinted from the author's "Bird Neighbors," which was written without a plan for a supplementary companion volume.

red; two middle tail feathers longest; all others banded with black and tipped with ashy white. Wing coverts sparsely spotted with black. Flanks and underneath the wings bluish.

Female—Duller and without iridescent reflections on neck.

Range—North America, from Quebec to Panama, and westward to Arizona. Most common in temperate climate, east of Rocky Mountains.

Season—March to November. Common summer resident; not migratory south of Virginia.

The beautiful, soft colored plumage of this incessant and rather melancholy love-maker is not on public exhibition. To see it we must trace the *a-coo-o, coo-o, coo-oo, coo-o* to its source in the thick foliage in some tree in an out-of-the-way corner of the farm, or to an evergreen near the edge of the woods. The slow, plaintive notes, more like a dirge than a love song, penetrate to a surprising distance. They may not always be the same lovers we hear from April to the end of summer, but surely the sound seems to indicate that they are. The dove is a shy bird, attached to its gentle and refined mate with a devotion that has passed into a proverb, but caring little or nothing for the society of other feathered friends, and very little for its own kind, unless after the nesting season has passed. In this respect it differs widely from its cousins, the wild pigeons, flocks of which, numbering many millions, are recorded by Wilson and other early writers before the days when netting these birds became so fatally profitable.

What the dove finds to ardently adore in the "shiftless housewife," as Mrs. Wright calls his lady-love, must pass the comprehension of the phœbe, which constructs such an exquisite home, or of a bustling, energetic Jenny Wren, that "looketh well to the ways of her household and eateth not the bread of idleness." She is a flabby, spineless bundle of flesh and pretty feathers, gentle and refined in manners, but slack and incompetent in all she does. Her nest consists of a few loose sticks, without rim or lining ; and when her two babies emerge from the white eggs, that somehow do not fall through or roll out of the rickety lattice, their tender little naked bodies must suffer from many bruises. We are almost inclined to blame the inconsiderate mother for allowing her offspring to enter the world unclothed—obviously not her fault, though she is capable of just such negligence. Fortunate are the baby doves when their lazy mother scatters her makeshift nest on top of one that a robin has deserted, as she frequently does. It is almost excusable to take her young birds and rear them in captivity, where they invariably thrive, mate, and live happily, unless death comes to one, when the other often refuses food and grieves its life away.

In the wild state, when the nesting season approaches, both birds make curious acrobatic flights above the tree-tops; then, after a short sail in midair, they return to their perch. This appears to be their only giddiness and frivolity, unless a dust-bath in the country road might be considered a dissipation.

In the autumn a few pairs of doves show slight gregarious tendencies, feeding amiably together in the grain fields and retiring to the same roost at sundown.

PART IV.

BIRDS OF PREY

Vultures
Kites
Hawks
Eagles
Owls

BIRDS OF PREY
(Order Raptores)

These rapacious birds, whose entire structure indicates strength, ferocity, carnivorous appetite, and powerful flight, have, for their diagnostic features, strong, hooked bills, covered toward the base with a cere, or membrane, through which the nostrils open; four long, strong toes, flexibly jointed to secure greatest grasping power, and fitted with sharp, curved nails or talons; long, ample wings and muscular legs, partly feathered. The young, though not naked when hatched, as are most altricial birds, remain in the nest, dependent on their parents, for a long time.

American Vultures
(Family Cathartidæ)

Two of the eight vultures found in the western hemisphere live chiefly in the southern part of the United States. These birds have head and neck bare of feathers or covered only with down; toes and tarsus bare likewise; claws not much incurved and not very sharp; perfectly developed wings for continuous, majestic flight; and strong digestive organs adapted to carrion, since these birds are most active scavengers. Vultures are gregarious. They alone, among the birds of prey, feed their young by disgorging food.

Turkey Vulture
Black Vulture

Kites, Hawks, Eagles
(Family Falconidæ)

A loud, startling cry; powerful legs and feet for striking at prey; the hind fourth toe as long as the others, for grasping;

sharp, decurved nails or talons, indicate the extreme of ferocity among the feathered tribes. Small mammals, reptiles, batrachians, and insects make up a far larger proportion of this family's food than birds and poultry, although agriculturists generally little appreciate its great service in protecting their crops. Solitary birds of freedom, they hold themselves high aloof from the world; nevertheless, eagerly vigilant, their wonderfully acute eyes keep constantly alert for food. Flocks are occasionally seen, but in the act of migrating only, for they are not truly gregarious, like vultures. Some species remain mated for life, and become strongly attached to a nesting site, where they return year after year, a pair preempting an entire neighborhood.

     Swallow-tailed Kite
     Marsh Hawk or Harrier
     Sharp-shinned Hawk
     Cooper's Hawk
     American Goshawk
     Red-tailed Hawk
     Red-shouldered Hawk
     Swainson's Hawk
     Broad-winged Hawk
     Rough-legged Hawk
     Golden Eagle
     Bald Eagle
     Duck Hawk
     Pigeon Hawk
     American Sparrow Hawk
     American Osprey, or Fish Hawk

Barn Owls

(Family Strigidæ)

A broad, triangular, facial disc; a jagged edged middle toe nail, and some peculiarities of bone structure, separate these birds from the other owls. They have also very long, pointed wings, reaching beyond the tail; soft, downy, speckled plumage; legs feathered to toes; extremely acute, long claws, and comparatively small eyes among other outer characteristics; but in habits they differ little from their kin.

     American Barn, or Monkey-faced Owl

Horned and Hoot Owls

(Family Bubonidæ)

Like the osprey in the hawk group, owls have a peculiarly flexible, reversible hind toe; eyes not capable of being rolled but set firmly in the sockets, necessitating the turning of the head to see in different directions; feathered discs around the eyes; loose, mottled plumage, some species with feathered ear tufts (horns), others without; hooked beaks and muscular feet for perching and for grasping prey:—these are their chief characteristics. Birds of the woodland, more rarely of grassy marshes and plains, nearly all nocturnal in habits, since their food consists mostly of small mammals that steal abroad at night to destroy the farmer's crops, the owls are among the most valuable of birds to the agriculturist. Unless too large, the prey is bolted entire—the hair, claws, bones, etc., being afterward ejected in matted pellets.

American Long-eared Owl
American Short-eared Owl
Barred or Hoot Owl
Saw-whet or Acadian Owl
Screech Owl
Great Horned Owl
Snowy Owl
American Hawk Owl
Burrowing Owl

AMERICAN VULTURES
(Family Cathartidæ)

Turkey Vulture
(Cathartes aura)

Called also: TURKEY BUZZARD

Length—30 inches; wing spread about 6 feet.

Male and Female—Blackish brown; wing coverts and linings grayish; head and neck naked and red, from livid crimson to pale cinnamon, and usually with white specks; base of bill red, and end dead white; feet flesh colored. Head of female covered with grayish brown, fur-like feathers. Young darker than adults; bill and skin of head dark and the latter downy. Nestlings of yellowish white.

Range—Temperate North America, from Atlantic to Pacific, rarely so far north as British Columbia; southward to Patagonia and Falkland Islands. Casual in New England.

Season—Permanent resident, except at extreme northern limit of range.

Floating high in air, with never a perceptible movement of its widespread wings, as it circles with majestic, unimpassioned grace in a great spiral, this common buzzard of our southern states suggests by its flight the very poetry of motion, while its terrestrial habits of scavenger are surely the very prose of existence. In the air the bird is unsurpassed for grace, as, rising with the wind, with only the slightest motion of its great, flexible, upturned wings, it sails around and around, for hours at a time, at a height of two or three hundred feet; then descending in a long sweep, rises again with the same calm, effortless soaring that often carries it beyond our sight through the thin, summer clouds. Humboldt recorded that not even the condor reaches greater heights beyond the summits of the Andes than this buzzard, which often joins its South American relative in its

TURKEY VULTURE.
½ Life size.

dizzy sport. Since the buzzard is gregarious, there are usually a dozen great birds amusing themselves by wheeling through space in pursuit of pleasure, and abandoning themselves to the amusement with tireless ecstasy. Is it not probable that so much exercise is taken to help digest the enormous amount of carrion bolted? For this reason, it is thought, the wood ibis soars and gyrates.

Other birds have utilitarian motives for keeping in the air; several of the hawks, for example, do indeed sail about in a similar graceful spiral flight, notably the red-tailed species, but a sudden swoop or dive proves that its slow gyrations were made with an eye directly fastened on a dinner. The crow soars to fight the hawk that carries off its young; the kingbird dashes upward to pursue the crow; but, amidst the quarrels and cruelties of other birds, the turkey buzzard sails serenely on its way, molested by none, since it attacks none, and makes no enemies, feeding as it does, for the most part, on carrion that none grudge it. The youngest chickens in the barnyard show no alarm when a turkey buzzard alights in their midst. They know that no more harmless creature exists. It is the most common bird in the South, being protected there by law in consideration of its service to the cities' street cleaning departments, which, in some places where Colonel Waring's methods are unheard of, it constitutes in the main. Every field has its buzzards soaring overhead and casting their shadows, like clouds, on the grain below. Depending on their services, the farmers allow the dead horse, or pig, or chicken to lie where it drops, for the vultures to peck at until the bones are as clean as if purified by an antiseptic. Fresh meat has no attractions for them; their preference is for flesh sufficiently fœtid to aid their sight in searching for food, and on such they will gorge until often unable to rise from the ground. When disturbed in the act of overhauling a rubbish heap in the environs of the city, for the bits of garbage that no goat would touch, they express displeasure at a greedy rival by blowing through the nose, making a low, hissing sound or grunt, the only noise they ever utter, and by lifting their wings in a threatening attitude. With both beak and claws capable of inflicting painful injury, the buzzard resorts to the loathsome trick of disgorging the foul contents of its stomach on an intruder.

Vultures

This automatic performance is practised even by the youngest fledglings when disturbed in the nest. It certainly is a most effective protection. Petrels also practise it, but not so commonly.

The turkey buzzard shows a decided preference for warm latitudes, never nesting farther north than New Jersey on the Atlantic coast, though, strangely enough, it has penetrated into the interior so far as British Columbia. Lewis and Clarke met it about the falls of the Oregon, and it is still not uncommon on the Pacific slope. Nevertheless, it is about the shambles of towns in the West Indies and other hot countries that the buzzard finds life the pleasantest. It has the tropical vice of laziness, so closely allied to cowardliness, and lives where there is the least possible necessity for exercising the stronger virtues. Our soldiers in the war with Spain tell of the final touch of horror given to the Cuban battle-fields where their wounded and dead comrades fell, by the gruesome vultures that often were the first to detect a corpse lying unseen among the tall grass.

As night approaches, one buzzard after another flies toward favorite perches in the trees, preferably dead ones, and settles, with much flapping of wings, on the middle branches; then stretching its body and walking along the roost like a turkey, until it arrives at the chosen spot, it hisses or grunts through its nostrils at the next arrival, whose additional weight frequently snaps the dead branch and compels a number of the great birds to repeat the prolonged process of settling to sleep. But, very frequently, the traveller in the South notices buzzards perched, like dark spectres, on the chimneys of houses, at night, especially in winter, in order to warm their sensitive bodies by the rising smoke, and, after a rain, they often spread their wings over the flues to dry their water-soaked feathers. This spread-eagle attitude is also taken, anywhere the bird happens to be, when the sun comes out after a drenching shower.

Without exerting themselves to form a nest, the buzzards seek out a secluded swamp, palmetto "scrub," sycamore grove, or steep and sunny hillside, and deposit from one to three eggs, usually two, in the cavity of a stump, or lay them directly on the ground, under a bush, or on a rock—anywhere, in fact, that necessity urges. Rotten wood is a favorite receptacle, but the angular bricks of ruined chimneys are not disdained. The eggs are of a dull yellowish white, irregularly blotched with chocolate brown markings, chiefly at the larger end. Very rarely eggs are found without these markings. Laying aside, for a time, his slothful ways, the male carefully attends his sitting mate. As a colony of buzzards, when nesting, indulges its offensive defensive action most relentlessly, few, except scientists, care to make a close study of the birds' nesting habits.

Black Vulture

(Catharista atrata)

Called also: CARRION CROW

Length—About 24 inches. Wing spread over four feet.
Male and Female—Dull black; under part of point of wings silvery gray; head, neck, and base of bill dusky; tip of bill and feet flesh colored or grayish; head and neck bare.
Range—Common in South Atlantic and Gulf states, through Mexico to South America. Occasional in Western states. Rare north of Ohio.
Season—Permanent resident.

With a heavier, more thickset body than the turkey buzzard's and shorter wings, this very common "carrion crow" may be identified in mid-air by its comparative lack of grace in flight, its frequent wing flapping, and its smaller size, which is more apparent than real, however, since its stocky build offsets its narrower wing-spread. Five or six quick, vigorous flaps of the wings send the bird sailing off horizontally; another series of wing flappings carries it up higher for another sail; but the flight is heavy and labored when compared with the majestic spiral floating of the buzzard, and it lacks the fascination that characterizes that other vulture's motion. Seen on the ground, the dusky head of the carrion crow is alone sufficient to differentiate it from the red-headed buzzard. It is also black instead of brown; and its tail is short and rounded.

A more southerly range and a decided preference for the seacoast, and for the habitations of men, again distinguish it; but in nesting and other habits than those noted these two vultures are almost identical. From North Carolina southward, every city and village contains a horde of these dusky scavengers, walking

Vultures

about the streets as familiarly as chickens to pick up the scraps of food that so quickly become putrid in a warm climate ; or, perched upon the chimney tops, drying and warming their grim, spectre-like bodies. Every market place is haunted by them more persistently than by the turkey buzzard ; for the carrion crows will be walked on by the crowd rather than leave the refuse of the butcher's stalls. One bird in Charleston, S. C., has visited a certain butcher regularly for twenty years. While both species are cowards, it is the black vulture that invariably secures the tidbit in the refuse heap from under the very beak of the turkey buzzard that stands in ridiculous awe of its heavy weight. But it is only at feeding time that these two vultures associate. The black vulture is decidedly the more gregarious. A carcass of horse or hog will sometimes be entirely concealed under an animate mass of these sable scavengers, perhaps two hundred or more fiercely clawing at the loathsome food. They gave the final touch of horror to the scene after the destruction of the Spanish fleet at Santiago when the sailors were washed ashore, and to the battlefields where our own dead soldiers lay. One of the Rough Riders who had shown magnificent courage in the presence of the enemy, went into violent hysterics at the sight of the vultures hovering over his fallen friends in the underbrush about Baiquiri.

KITES, HAWKS, EAGLES, ETC.
(Family Falconidæ)

Swallow-tailed Kite
(Elanoides forficatus)

Called also: FORK-TAILED KITE ; SNAKE HAWK.

Length—About 24 inches, or according to development of tail. Wing spread about 4 feet.

Male and Female—Head, neck, under parts, including wing linings, band across lower back, snow white ; rest of plumage glossy black, showing violet and green reflections. Bill bluish black; feet and very short legs, light. Tail 14 inches long and cleft like a swallow's for half its length.

Range—United States, especially in the interior, from Pennsylvania and the great plains southward to Central and South America. Casual in New England, Minnesota, Manitoba, and Assiniboia; nesting irregularly throughout its range ; winters chiefly south of United States.

Season—Summer resident. April to October.

Not excepting even the turkey vulture, the tern or the swallow, no bird moves through the sky with more exquisite grace and buoyancy than this beautiful black and white, sharp winged kite, whose motion combines the special fascinations of each of its three close rivals. Soaring upward, buzzard fashion, until it sometimes fades from sight, or floating like it on motionless pinions; now swooping with the dash of a tern and catching itself suddenly just above the earth to skim along the surface like a swallow; swaying its trim body with a cut of the wing and the lashing of its long forked tail, it pauses neither for rest nor food, but apparently spends every waking moment in the air. It is supposed it even sleeps while it floats, so little conscious effort is evident in its flight; and it feeds a-wing by tearing off bits of the snake, or other prey, firmly grasped in its small feet. This has been seized while passing and without pause. In this way too the bird takes a drink. Because they are so little used for walking, for one almost never sees this kite on the ground, its legs are very short and all but invisible.

Most abundant in the western division of the Gulf states and above the great plains, the numbers of this bird—let it be recorded—nowhere seem to have diminished, since it feeds almost exclusively on snakes, lizards, and the larger insects such as locusts, crickets, and grasshoppers, and never on other birds. Even the dullest mind recognizes it as harmless and beneficent. Naturally a bird so little persecuted shows no great fear of man. Its shrill, penetrating *wee-wee-wee* has been uttered in the very ears of a picnic party within sight of a huge hotel in Minnesota.

But when the nesting season arrives, these kites seek out uninhabited, inaccessible regions where it is well worth while to follow them, however, since their flight, always charming, dashing, and elegant, now assumes matchless perfection impossible to describe. Even their wooing is done on the wing. Several pairs may build in a neighborhood, which is usually a dense wood near water that attracts their prey within easy reach; and at the top of some tall, straight tree, anywhere from sixty to one hundred and forty feet from the ground, an irregular nest of large loose twigs, lined or unlined with moss, may likely as not rise from the foundations of one used the previous year. From two to four white eggs, boldly spotted or blotched with different shades of brown, are laid any time from April to June, according to the latitude. It is thought both kites take turns at the incubating, which is closely attended to; or at least the male is particularly devoted to his sitting mate, always being seen near by. In leaving the nest a bird rises upward suddenly as if sent up by a spring, instead of flying sidewise as most birds do; and in alighting it first poises itself directly above the eggs, then descends on apparently motionless wings so softly and lightly the large body might be a single feather dropping from the sky.

Marsh Hawk

(Circus Hudsonius)

Called also: MARSH HARRIER; BLUE HAWK; MOUSE HAWK

Length—Male 19 inches; female 22 inches.

Male—Upper parts gray or bluish ash, washed with brownish; upper tail coverts pure white; silver gray tail feathers with five or six dusky bars, the outer primaries darkest; upper breast pearl gray, shading into white underneath, where the plumage is sparsely spotted with rufous. Hooked bill, and feet black.

Female and Young—Upper parts dark amber; the head and neck streaked, other parts margined or spotted with reddish brown; upper tail coverts white; middle tail feathers barred with gray and black, others barred with pale yellow and black. Under parts rusty buff, widely streaked on breast and more narrowly underneath with dusky. The younger the bird the heavier its blackish and rufous coloration, many phases of plumage being shown before emerging into the gray and white adult males.

Range—North America in general, to Panama and Cuba; nests throughout North American range; winters in southern half of it.

Season—Summer resident at northern half of range.

Close along the ground skims the marsh hawk, since field mice and other small mammals, frogs, and the larger insects that hide among the grass are what it is ever seeking as it swerves this way and that, turns, goes over its course, "quartering" the ground like a well trained dog on the scent of a hare—the peculiarity of flight that has earned it the hare-hound or harrier's name. A few easy strokes in succession, then a graceful sail on motionless wings, make its flight appear leisurely, even slow and spiritless, as compared with the impetuous dash of a hawk that pursues feathered game; hence this is counted an "ignoble" hawk in the scornful eyes of falconers used to the noble sport of hawking. Open stretches of country, wide fields, salt and fresh water marshes, ponds, and the banks of small streams, whose sides are not thickly wooded, since trees simply impede this low flier's progress, are its favorite hunting grounds; and it will sometimes alight on a low stump, or in the grass itself, for it is a low percher too. Because its quarry is humble, and farmers, on the whole, appreciate its service in destroying meadow mice, crickets, grasshoppers, and other pests, this bird suffers comparatively little persecution, and still remains one of the most widely distributed and common of its tribe. That it occasionally preys upon small birds, when other food fails, cannot be denied; but nearly one-half of all the stomachs examined by Mr. Fisher, for the Department of Agriculture, contained mice.

In the nesting season especially, the harrier belies that name, but, proving his title to *Circus*, his Latin one, wheels round and round and floats high above the earth, describing some beautiful evolutions as he goes, that are calculated simply to stimulate afresh the ardor of his well beloved, since evidence strongly points to a life partnership between the mates. Soaring in the sky, suddenly he falls, turning several somersaults in the descent. "At other times," says Mr. Ernest Seton Thompson," he flies across the marsh in a course which would outline a gigantic saw, each of the descending parts being done in a somersault, and accompanied by screeching notes which form the only love song within the range of his limited powers." All hawks have a screaming, harsh cry, not distinctly different in the different species to serve as a clew to identity except to those well up in field practice; but the white lower back of the harrier, its long tail, and its terrestrial habits serve to identify it in any phase of plumage. Owing to its long wings, it appears much larger in the air than on the ground. Four to six dull or bluish white eggs, unmarked, are laid in May, in a nest built of twigs, hay, and weeds, on the ground; yet the clumsy affair was the joint effort of the mates, that also take turns in sitting and in feeding the young.

Sharp-shinned Hawk

(Accipiter velox)

Called also: PIGEON HAWK; LITTLE BLUE DARTER

Length—Male 10 to 12 inches; female 12 to 14 inches.
Male and Female—Upper parts slaty gray. Tail, which is about 3 inches longer than tips of wings and nearly square, is ashy

MARSH HAWK.

gray, barred with blackish, and with a whitish tip; throat white, streaked with blackish. Other under parts whitish, barred on sides and breast with rusty, buff, and brown, lining of wings white, spotted with dusky; head small; tarsus slender and feathered half way; feet slender. Immature birds have dusky upper parts, margined with rufous; tail resembling adults'. Under parts buff or whitish, streaked or spotted with rusty or blackish.

Range—North America in general; nesting throughout the United States and wintering from Massachusetts to Guatemala.

Season—Permanent resident, except at northern parts of range.

A smaller edition of Cooper's hawk (to be distinguished from it chiefly by its square, instead of rounded, tail), like it, dashes through the air with a speed and audacity that spread consternation among the little song and game birds and poultry, once it appears, like a flash of "feathered lightning," in their midst. Cries of terror from many sympathizers when a sparrow, a goldfinch, a warbler, or some tiny victim is making desperate efforts to escape, first attract one's notice; but of what avail are the stones hurled after a hawk that swoops and dodges, twists and turns, in imitation of every movement of the panic stricken bird he presses after, closer and closer, until, at the end of a long chase, when it is exhausted and almost worried to death, he strikes it with talons so sharp and long that they penetrate to the very vitals? Now alighting on the ground, he rends the warm flesh from its bones with a beak as savage as the talons. If the little bird had but known enough to remain in the thicket! A race for life in the open seems to give the pursuing villain a fiendish satisfaction: let his little prey but dash toward the woods, where he knows as well as it does that it is safe, and one fell swoop cuts the journey short. There can be little said in praise of a marauder that boldly enters the poultry yard and devours dozens of chicks, attacks and worsts game birds quite as large as itself, and that eats very few mice and insects and an overwhelming proportion of birds of the greatest value and charm. The so called "hen-hawks" and "chicken-hawks"—much slandered birds—do not begin to be so destructive as this little reprobate that, like its larger prototype and the equally villainous goshawk, too often escape the charge of shot they so richly deserve.

Unhappily, the sharp-shinned hawk is one of the most abundant species we have. Doubtless because it is small and looks inoffensive enough, as it soars in narrow circles overhead, its worse than useless life is often spared.

Cac, cac, cac, very much like one of the flicker's calls, is this hawk's love song apparently, for it seldom, if ever, lifts its voice, except at the nesting season. Now it seeks the woods to make a fairly well constructed nest of twigs, lined with smaller ones, or strips of bark, with the help of its larger mate, from fifteen to forty feet from the ground. Strangely enough, the nest is not a common find, however abundant the bird, neither Nuttall nor Wilson having discovered one in all their tireless wanderings. Dense evergreens, the favorite nesting localities, conceal the nest, large as it is—much too large for so small a bird, one would think. A pair of these hawks may sometimes repair their last season's home, but will never appropriate an old tenement belonging to others, as many hawks do. Late in May, or even so late as June, from three to six bluish or greenish white eggs, heavily blotched or washed with cinnamon red or chocolate brown, keep both parents busy incubating and, later, feeding a hungry family. Climb up to the nursery, and angry, fearless birds dash and strike at an intruder as if he were no larger than a goldfinch.

Cooper's Hawk
(Accipiter Cooperi)

Called also: CHICKEN HAWK; BIG BLUE DARTER.

Length—Male 15.50 inches; female 19 inches.
Male, Female and Young—To be distinguished from the sharp-shinned species only by their larger size, darker, blackish crowns, and rounded, instead of square, tails.
Range—Temperate North America, nesting throughout its United States range; some birds wintering in Mexico and the southern states.
Season—Permanent resident except at northern limits of range, where it is a summer or transient visitor.

Like the sharp-shinned hawk in habits as in plumage, this, its larger double, lives by devouring birds of so much greater value than itself that the law of the survival of the fittest should be enforced by lead until these villains, from being the commonest of their generally useful tribe, adorn museum cases only. Captain Bendire, writing for the Government, says: "Cooper's hawk must be considered as one of the few really injurious Raptores found within our limits, and as it is fairly common at all seasons throughout the greater part of the United States, it does in the aggregate far more harm than all other hawks. It is well known to be the most audacious robber the farmer has to contend with in the protection of his poultry, and is the equal in every way, both in spirit and dash, as well as in bloodthirstiness, of its larger relative, the goshawk, lacking, however, the strength of the latter, owing to its much smaller size. It is by far the worst enemy of all the smaller game birds, living to a great extent on them as well as on small birds generally. It does not appear to be especially fond of the smaller rodents; these, as well as reptiles, batrachians, and insects, seem to enter only to a limited extent into its daily bill of fare, and unfortunately it is only too often the case that many of our harmless and really beneficial hawks have to suffer for the depredations of these daring thieves."

American Goshawk

(Accipiter atricapillus)

Called also: **BLUE HEN HAWK; PARTRIDGE HAWK.**

Length—Male 22 inches; female 24 inches.

Male and Female—Upper parts bluish slate, darkest or blackish on head; white line over and behind eye; tail like back and banded with blackish bars, the last one the broadest, and the tip whitish. Entire under parts evenly marked with irregular wavy lines of gray and white, the barring usually most heavy on the flanks and underneath. Immature birds have dusky upper parts margined with chestnut, the tail brownish gray barred with black, the under parts white or buff streaked with black. Bill dark bluish. Feet yellow.

Range—Northern North America; nests from northern United States northward; winters so far south as Virginia.

Season—Permanent resident.

Another villain of deepest dye; what good can be said of it beyond that it wears handsome feathers, is a devoted mate and parent, a fearless hunter, and of some small, if disproportionate, value to the farmer in occasionally eating field mice and insects? Whitewashing is useless in the case of a bird known to be the most destructive creature on wings. No more daring marauder prowls above the poultry yards than the goshawk that drops like a thunderbolt from a clear sky at the farmer's very feet and carries off his chickens before his eyes. Grouse, Bob Whites, ducks, and rabbits:—in fact, all the sportsmen's pets and innumerable songbirds, are hunted down with a dash and spirit worthy of a better motive. Bloodthirsty, delighting in killing what it often cannot eat, marvellously keen sighted, a powerful, swift flyer, aggressive, and constantly on the alert, it is small wonder all lesser birds become panic-stricken when this murderer sails within striking distance. Without a quiver of its wings it will sail and sail, apparently with the most innocent intent. Again, with strong wing beats, it will rush through the air and overtake a duck that flies at the rate of a hundred miles an hour, seize it by the throat, sever its windpipe and fly off with its burden. One very rarely sees the goshawk perching and waiting for prey to come to it. When it does so, it holds itself erect, elegant and spirited as ever. After tearing the legs off a ruffed grouse, and plucking every feather, this villain has been known to prepare another and another until five were ready for an orgie, which consisted of only fragments of each, torn with its savage beak. Mr. H. D. Minot tells of watching a goshawk press into a company of pine grosbeaks and seize one in each foot. Happily the agony is short, for a hawk's talons penetrate the vitals.

Although a northern ranger, the goshawk nests early—in April or early May—and placing a quantity of twigs and grasses close to the trunk of a tree, anywhere from fifteen to seventy feet from the ground, both mates take turns in attending to the nursery duties after from two to four pale bluish green eggs (that fade to dull white) have been laid. Now the hawks are more audacious and vicious than ever, as their piercing cries indicate, and it is an irrepressible collector who dares rob them.

Red-tailed Hawk

(Buteo borealis)

Called also: HEN HAWK; CHICKEN HAWK; RED-TAILED BUZZARD; RED HAWK

Length—Male 20 inches; female 23 inches.

Male and Female—Upper parts dark grayish brown; the feathers edged with rufous, white, gray, and tawny; the wing coverts lack the rufous shade; tail rusty red, tipped with white and with a narrow black band near its end, but silvery gray on the under side. Under parts buff or whitish, with heaviest brown or blackish markings on the flanks and underneath, often forming an imperfect band across the lower breast. Immature birds lack the red tail, their tails being grayish, or like the back, with numerous black bars.

Range—Eastern United States, west to the great plains; nesting throughout its range.

Season—Permanent resident; partly migratory.

With a wing spread of four feet, the red-tailed hawk, no less than the red-shouldered species, is a conspicuous object in the sky, especially in August and September, when all hawks appear to be less hungry and vicious than usual, and constantly and serenely sailing and gyrating high overhead, beyond thought of mundane concerns. Lacking the dash and address of Cooper's hawk, this far larger, heavier buzzard is rather leisurely, not to say slow, of movement. Mounting higher and higher in a spiral till it appears a mere speck in the blue, it will sail and float, ascending, descending, in long undulations, then, when rising and circling, with no perceptible vibration of its wings, it will suddenly lift them to a point above the back and shoot earthward like a meteor. Catching itself just as you believe it must certainly dash itself to pieces, again it rises, with bounds, on broad wings to enjoy the stratum of cooler air, high above the tree tops, all these hardy birds delight in. One hawk was watched in the air, without once alighting, from seven in the morning till four in the afternoon.

When not in the act of sailing, the most likely position to find this majestic air king in is perched on a tree at the edge of a patch of woods, a dead limb near water, or above low open fields or swamps, and there, intent and eager, it will wait hours and hours for its quarry to come within range. Then, **like** feathered lightning, down it flashes and strikes its prey. One never sees this hawk dashing through the air **in** pursuit of a victim, **as the** sharp-shinned, Cooper's hawk and the goshawk do. It **may** sometimes **pounce** upon a bird a-wing, but humbler quarry generally takes it to earth. Of **the five** hundred and sixty-two stomachs **of red-tailed hawks examined by Mr.** Fisher for the **Department of Agriculture, one-half contained mice,** about one-third other mammals, fifty-four contained poultry or game birds; and batrachians or reptiles, insects, etc., filled part of the remainder, eighty-nine being empty. Captain Bendire, in his valuable book prepared for the Government, says : "Unfortunately the red-tailed hawk has a far worse reputation with the average farmer than it really deserves; granting that it does capture a chicken or one of the smaller game birds now and then— and this seems to be the case only in winter, when such food as they usually subsist on is scarce—it can be readily proved that it is far more beneficial than otherwise, and really deserves protection, instead of having a **bounty placed** on its head, **as has been** the case in several states."

Around the nest especially, though one sometimes hears its squealing whistle, like "escaping steam," as it floats overhead, at any season, the red-tail becomes more noisy, but its voice is rather weak, considering the size of the bird. About eighty per cent. of all nests found have been in birch trees, and placed from sixty to seventy feet from the ground. A large bundle **of sticks,** lined with strips of bark, twigs, and feathers from **the birds** themselves, is placed usually where some large limb **branches off** from the trunk; and so dear does this rude cradle become to **the mates that** jointly prepare it, it will be **used** year after year if the hawks are unmolested. From **two to four** dull white eggs, with rough, granulated shells, **often** scantily and irregularly marked with shades of cinnamon, take about four weeks of close incubation, in which both the **devoted** lovers and parents assist. It is believed these birds, **like most of** their kin, remain mated for life. The helpless, downy young remain in the nest until fully able to fly. Hawks usually bolt their food, and around a **nest are abundant traces of the** hearty appetite of a young family, the tufts of mouse hair and pellets of other disgorged, indigestible material plentifully besprinkling the ground.

. . . .

The Western Red-tail *(Buteo borealis calurus)*, a darker colored race than the preceding, differs from it in no essential particulars.

Red-shouldered Hawk

(Buteo borealis)

Called also: HEN HAWK; CHICKEN HAWK; WINTER HAWK; WINTER FALCON; RED-SHOULDERED BUZZARD.

Length—Male 18 to 20 inches; female 20 to 22 inches.

Male and Female—Rich dark reddish brown above, the feathers more or less edged with rufous, buff and whitish; lesser wing coverts rusty red, forming a conspicuous patch on shoulders; four outer feathers of wings notched and all barred with black and white; tail dark with white bars; under parts rusty or buff, the throat streaked with blackish, elsewhere irregularly barred with white; feet and nostrils yellow. Immature birds plain dark brown above, the wing patch sometimes indicated, sometimes not; head, neck, and under parts pale buff, fully streaked with dark brown; wing and tail quills crossed with many light and dark bars.

Range—Eastern North America from Manitoba and Nova Scotia to the Gulf states and Mexico, westward to Texas and the great plains; nests throughout its range.

Season—Permanent resident.

To shoot this commonest of the hawks has long been regarded as a virtue among farmers in the unfounded belief that it is an enemy to their prosperity; but the Department of Agriculture has prepared a special bulletin on the hawks and owls for their enlightenment, and the two so-called "hen hawks" have proved to be among the most valuable allies the farmer has. Of two hundred and twenty stomachs of the red-shouldered hawk examined by Mr. Fisher, only three contained remains of poultry; one hundred and two contained mice; ninety-two insects; forty, moles and other small mammals; thirty-nine, batrachians; twenty, reptiles; sixteen, spiders; twelve, birds; seven, crawfish; three, fish; two, offal; one, earthworms; and fourteen were empty. Let the guns be turned toward those bloodthirsty, audacious miscreants, Cooper's, the sharp-shinned hawk, and the goshawk, and away from the red-tailed and red-shouldered species, beneficent, majestic kings of the air! Longfellow, in "The Birds of Killingworth," among the "Tales of a Wayside Inn," has written a defence of the hawks, among other birds, that the Audubon societies might well use as a tract.

Sailing in wide circles overhead like the larger red-tail, the red-shouldered buzzard is a picture of repose in motion. Rising, falling in long undulations, floating, balancing in a strong current of the cool stratum of air far above the earth all this hardy tribe delight in, now stationary on motionless wings, and again with a superb swoop a very meteor for speed, the flight of this hawk has been familiar to us all from childhood, yet who ever tires watching its fascinating grace? Serenely the hawk pursues its way, ignoring the impudence of the small kingbird in pursuit and the indignities of the crow that may not reach the dizzy heights toward which it soars in wide spirals. While the mates are nesting from April to August, the helpless fledglings give them little opportunity to enjoy these leisurely sails; but toward the end of August, particularly in September, and throughout the winter, they are birds of freedom indeed. *Kee you, kee you*, they scream as they sail—a cry the blue jay out of pure mischief has learned to imitate to perfection. It is the red-tail, however, that screams most a-wing.

"Toward man the 'hen hawks' are naturally shy," says Minot; "but it is generally easy to approach them when gorged, or at other times to do so in a vehicle or on horseback. On a horse I have actually passed under one. They frequently leave their food when approached, instead of carrying it off in the manner of many hawks. Like other barbarians, they refuse to show signs of suffering, or to allow their spirit to become subdued. When shot and mortally wounded they usually sail on unconcernedly while their strength lasts, until obliged to fall. If not dead, they turn upon their rump, and fight till the last, like others of their tribe. Their eyes gleam savagely and they defend themselves with both bill and talons. With these latter, if incautiously treated, they can inflict severe wounds, and they sometimes seize a stick with such tenacity that I have seen one carried half a mile through his persistent grasp."

RED SHOULDERED HAWK.
½ Life-size.

The red-shouldered hawk spends most of its life perching, usually on some distended dead limb where, like an eagle in its dignity, it watches for mice and moles to creep through the meadow, chipmunks to run along stone walls, gophers and young rabbits to play about the edges of woods, frogs, snakes, etc., to move along the sluggish streams of low woodlands, its favorite hunting grounds. It is not shy, and when it perches may be quite closely approached and watched as it descends like a thunderbolt to strike its humble quarry, that is usually borne aloft to be devoured piecemeal. One never sees this hawk chasing a bird through the air as the tyrannical Cooper's hawk does. In nesting habits there is no noteworthy difference from the red-tails', beyond that the eggs are a trifle smaller.

. . . .

Swainson's Hawk or Buzzard (*Buteo Swainsoni*), an infrequent visitor east of the Mississippi, is nevertheless the commonest of all its tribe in some sections of the West. In the many phases of plumage shown between infancy and old age, this large, amiable fellow may always be distinguished by the three notched outer primaries of his wings taken in connection with his size, about twenty inches, and his dusky brown upper parts more or less margined with rufous or buff; the unbarred primaries of wings; his grayish tail indistinctly barred with blackish, which shows more plainly from the under side; the large rusty patch on his breast, and by the white or buff under parts that are streaked, spotted, or barred with blackish, rusty, or buff. Preeminently a prairie bird, it prefers the watercourses of lowlands that are scantily timbered and the cultivated fields for hunting grounds, since mice, gophers, frogs, grasshoppers, crickets, and such fare —rarely if ever a bird—are what it is ever seeking. Therefore from the most selfish of economic standpoints it should enjoy the fullest protection. Gentle, unsuspicious, living on excellent terms with its humblest feathered neighbor, mated for life to its larger spouse, and an unselfish, devoted parent, Swainson's hawk has more than the average number of virtues to commend it to mankind.

Broad-winged Hawk

(Buteo latissimus)

Length—Male 14 inches; female 16 inches.

Male and Female—Upper parts dusky grayish brown more or less bordered with rusty and buff; blackish tail with two bars and the tip grayish white; three outer primaries of wings notched; under parts heavily barred with white or buff and dull chestnut brown, the dark in excess on the front parts, the white predominant underneath; most of the feathers black shafted, giving the effect of pencilling, particularly on white throat; wing linings white with some reddish or blackish spotting.

Range—Eastern North America from New Brunswick and the Saskatchewan to the Gulf of Mexico, northern South America and the West Indies; nests throughout its United States range.

Season—Summer resident. May to October.

This is the hawk of the Adirondacks among other favorite resorts, and since it comes north chiefly to nest, no place is too inaccessible for it to seek out, no retreat too lonely for these devoted mates, that ever delight most of all in each other's company. While its range is wide, it is locally common in a few places and rare in others, a lover of wild, unvisited regions while it has serious concerns to attend to, and only during the spring and autumn migrations, therefore is it much in evidence; but nowhere and at no time so common about farms and the habitations of men as the red-tailed and the red-shouldered "chicken hawks" that, on the contrary, have nothing to do with mountain fastnesses.

Yet the broad-winged species is perhaps the least suspicious and approachable hawk we have; gentle and never offering to strike at an intruder no matter in what distress of mind concerning its nest; inoffensive to its smallest feathered neighbor; lacking in the spirit and dash of a Cooper's hawk, and also in that murderer's bloodthirstiness; and quiet except just near its home. There one sometimes hears the *chee-e-e-e* of one mate sitting on some distended dead limb, answered by the other lover for hours at a time during the nesting season. Like most of its tribe, both mates construct a bulky nest of twigs high in some tree close to the trunk, and, if necessary, will repair an old nest from year to year rather than leave a beloved home. From two to four dull or buff white eggs spotted, blotched, or washed with yellow or cinnamon brown, keep both parents closely confined by turns during the four weeks of early summer that must elapse before the downy helpless fledglings begin to clamor for grasshoppers, beetles, crickets, mice, gophers, squirrels, shrews, small snakes and frogs (very rarely small birds), that must be consumed in large quantities judging from the quantity of pellets of hair and other indigestible material found below the cradle. The farmer has every reason to protect so valuable an ally.

Although it appears sluggish, and even stupid, when perching after a gorge, the broad-winged hawk naturally would be a graceful, easy flyer. Gliding through the air in spirals so high that one sometimes loses sight of its heavy, broad body, it has been seen swooping suddenly to earth, like a meteor; then catching itself before dashing its body to pieces on the mountain side, it will fly off, with short, rapid strokes, at high speed.

The Rough-legged Hawk *(Archibuteo lagopus sancti-johannis)*—the hare-footed hawk of St. John, New Brunswick—is almost too variable in plumage to be briefly described, but whether in its dark, almost blackish, phase, when it is known as the black hawk; or in the light phase, when its dusky upper parts are mixed with much white and buff, and its whitish under parts are streaked and spotted with black to form a band across the lower chest, it may always be known by its fully feathered legs. In the United States it is chiefly a spring and autumn migrant, or a winter visitor, for it goes to the fur countries to nest. The material for a cradle, usually placed on a cliff, would fill a wheelbarrow. Its range is over the whole United States, Alaska, and the British possessions. One occasionally meets this large, heavy prowler at the dusk of evening, when mice and the other small rodents, crickets and such humble quarry creep timidly forth, flying with noiseless, measured, owl-like pace, quite low along the ground, like the harrier, and ready to pounce upon a victim. Or again, it may be sitting on a low branch, sluggishly waiting for its prey to come within striking distance. Its choice of food is calculated to win for the hawk the friendship of the intelligent farmer.

Golden Eagle

(Aquila chrysaetos)

Called also: RING-TAILED EAGLE; MOUNTAIN EAGLE; WAR EAGLE

Length—Male 30 to 35 inches; female 5 inches longer.

Male and Female—Back of the head and nape pale yellow; lower two-thirds of tail white, leaving a broad, dark band across end; legs entirely feathered with white; rest of plumage dusky brown. Immature birds are similar, darker; base of tail has broken grayish bars, and feathers on legs and under tail coverts are buff. Perfect plumage not developed under three years. Birds "grow gray" with age.

Range—North America, south to Mexico; nesting within United States, Europe, and Northern Asia.

Season—Permanent resident.

> "He clasps the crag with hooked hands;
> Close to the sun, in lonely lands,
> Ringed with the azure world he stands.
> The wrinkled sea beneath him crawls;
> He watches from his mountain walls,
> And like a thunderbolt he falls."
>
> —*Tennyson.*

Restricted chiefly to the mountainous parts of unsettled regions on three continents, this magnificent bird is best known on this one to the Indians, who write no bird books, however. It is their emblem for whatever is courageous, fierce, and successful in war. All birds of prey typify these qualities to the Indian mind, it is true, but the eagle stands at the lead; its feathers, fit head-dress for any chief, were chosen to inspire him with the bird's prowess. The buzzards and eagles represent their old men—those sages who have little hair, or those whose locks are white—hence to these birds have the secrets and the wisdom of ages been confided, and a respect akin to worship is shown them. What the imperial eagle meant to the terrorized ancients we can little guess in these days of democratic ideals. From the bird's majestic soaring, what more natural than to suppose it communicated directly with the gods on Mount Olympus, and was Jove's favored messenger? Certain Asiatic tribes believe that arrows plumed with eagle quills certainly reach the heart of an enemy; but what connection there may or may not be between these beliefs and those of our redskins no ethnologist has said.

Larger than the European golden eagle, and in every way "better," our golden eagle "is a clean, trim-looking, handsome bird," says Captain Bendire, "keen sighted, rather shy and wary at all times, even in thinly settled parts of the country, swift of flight, strong and powerful of body, and more than a match for any animal of similar size. In the West, where food is still plenty, their bill of fare is quite varied. This, I am informed, includes, occasionally, young fawns of antelope and deer, but more frequently small mammals of different kinds, as the yellow-bellied marmot, prairie dogs, hares, wood rats, squirrels, and smaller rodents, water fowl, from wild geese to the smaller ducks and waders, grouse and sage fowl. On the extensive sheep ranches they are said to be occasionally quite destructive to young lambs." Several seemingly well authenticated cases of the golden eagle carrying off very young children are recorded in this country and Europe, but our authorities sneer at them.

Strangely enough, a pair of eagles, instead of being fiercely aggressive, as one would suppose, when their nest is approached, are quite indifferent and will circle around at a great height and watch the intruder with unimpassioned calm, or else entirely disappear. Trees or rocky cliffs seem to be chosen for nesting sites indiscriminately, the abundance of food in any vicinity being their first consideration in the choice of a home. Each pair of eagles have their fixed range of five or six miles, or more, and become so attached to it only persistent persecution will drive them away. Some nests are quite five feet in diameter, and contain twigs, weeds, hay, cattle hair, and feathers enough to fill a wagon; others are no larger than a hen hawk's; nearly all are flat on top, with just enough depression to bring the top of the egg on a level with the side. For a few days before the eggs are laid, a pair of eagles will perch, side by side, hours at a time, an attitude common to many birds of prey at this tender season. Two or three dull white, roughly granulated eggs, sometimes plain, more often blotched or speckled with brown, appear at an interval of two or three days, or even a week; after four weeks of constant incubating by both parents, the fluffy white brood appear; but, although they grow rapidly, it is fully two months before they leave the eyrie. Just as soon as they can fly and secure a living, the old birds cast them off. They are three years in perfecting their plumage, it is said, and they may live a century.

Bald Eagle

(Haliætus leucocephalus)

Called also: WHITE-HEADED EAGLE; WASHINGTON EAGLE; AMERICAN EAGLE; BALD SEA EAGLE.

Length—Male 30 to 33 inches; female 35 to 40 inches.

Male and Female—Head, neck, and tail white; after third year rest of plumage dusky brown, the feathers paler on edges; bill and feet yellow; legs bare of feathers. Immature birds are almost black the first year ("black eagles"); the bases of feathers white; bill black. Second year they are "gray eagles" and are then actually larger than adults. The third year, they come into possession of "bald" heads and white tails.

Range—North America, nesting throughout range.

Season—Permanent resident.

Emblem of the republic, standing for freedom to enjoy life, liberty, and the pursuit of happiness, it must be owned that our national bird is a piratical parasite whenever he gets the chance. "The majority of the *Falconidæ* have an attractive physique and superior strength as well as a haughty bearing," says Mr. Chamberlain. "They are handsome, stalwart ruffians, but they are nothing more. They are neither the most intelligent nor most enterprising of birds, nor the bravest. They are not even the swiftest or most dexterous on the wing; and in bearing, proudly as they carry themselves, are not supreme." With every provision of nature for noble deeds: keenest sight, superb strength, hardihood, fully developed wings, it is seldom that the American eagle obtains a bite to eat in a legitimate way, but almost invariably by stratagem and plunder. Near the sea and other large bodies of water he sits in majesty upon a cliff, or on the naked limb of some tree commanding a wide view, and watches the osprey—a conspicuous sufferer—and other water fowl course patiently over the waves up and down the coast for a fish. Instantly one is caught, down falls the eagle like Jove's thunderbolt from Mount Olympus, and as escape from so overpowering a foe is impossible, the successful fisher quickly drops its prey, while the eagle, dexterously catching it before it touches the water, makes off to his eyrie among the clouds to enjoy it at leisure. Dead fish cast up on the beach, carrion disgorged by intimidated vultures, sea and shore birds (particularly in the South) are devoured by this rapacious feeder. Ducks, geese, gulls, and notably coots, that he condescends to catch himself, are favorite morsels when fish fail. It is said wounded birds suit this unsportsmanlike hunter best. These are picked clean of feathers before the flesh is torn from their bones. In the interior young domestic animals are carried off, but scientists raise their eyebrows at tales of children being borne away by eagles; yet it would seem that some rare instances are well authenticated. Audubon had an adult male in captivity that weighed only fourteen and a half pounds, and although it ate enormously one may grant that an uncaged bird might weigh twenty pounds; still a young child often exceeds that figure, and there is the great resistance of the air to be overcome as well.

When the nesting season approaches, which in the south begins in February and at the far north in May, the eagles may be seen hunting in couples and soaring in great spirals with majestic calm at a dizzy height. As they swoop earthward, the tops of the trees over which they pass sway in the current of air they create. These birds, like most of their class, remain mated throughout their long life, but often quarrel at other seasons than this, when one encroaches upon the prescribed territory where the other is hunting. Now they are especially noisy: *cac-cac-cac* screams the male, a sound too like a maniac's laugh to be pleasant. The cry of the female is more harsh and broken, sufficiently different for one well up in field practice to tell the sex of the bird by its voice.

A tall pine tree near water is, of all nesting sites, the favorite. Next to that a rocky ledge of some bold, inaccessible cliff, or that failing too, the bulky cradle may be laid directly on the ground; but whatever site may be chosen, that forever remains home, a shelter at all seasons, the dearest spot on earth. An immense accumulation of sticks, sod, weeds, corn stalks, hay, pine tops, moss, and other coarse materials make a flat structure four or five feet in breadth and sometimes of even greater height after a succession of annual repairs. While the two or three large, rough, dull white eggs are being incubated by both mates, and especially after the young appear, these eagles, unlike the golden species, become truly magnificent in the fierce defence of their treasures; yet a rooster is easily a match for the cowardly eagle at other times. Immense quantities of food must be carried to the helpless young for the three or four months while they remain in the nest, and for weeks after they learn to fly. Because immature birds reverse nature's order and are larger than adults, and their plumage undergoes three changes before they appear at the close of the third year in white heads and tails, some early writers described the black eagle, Washington's eagle, and the bald-headed eagle as three distinct birds, even Audubon and Nuttall treating this one species as two. In whatever phase of plumage, one may know our national bird by its unfeathered tarsi. It is safe to say any eagle seen in the eastern United States is the bald-head, which name, of course, does not indicate that the bird is actually bald like the vultures, but simply hooded with white feathers.

Duck Hawk

(Falco peregrinus anatum)

Called also : PEREGRINE, WANDERING, MOUNTAIN, OR ROCK FALCON ; GREAT-FOOTED HAWK

Length—Male, 16 inches ; female, 19 inches.
Male and Female—Upper parts dark bluish ash, the edges of feathers paler ; under parts varying from dull tawny to whitish, barred and spotted with black, except on throat and breast. A black patch on each cheek gives appearance of moustache. Wings stiff, long, thin, and pointed. Tail and upper coverts regularly barred with blackish and ashy gray. Bill bluish, toothed, notched ; the cere yellow. Talons long and black.
Range—North America at large. South to Chile. Nests locally throughout its United States range.
Season—Chiefly a winter visitor, but a perpetual rover.

The falconers of Europe divided birds employed in their sport into two classes, those of falconry and those of hawking; the latter class containing such "ignoble" birds as our goshawk, broad winged buzzard, the sparrow hawk, and those of their kin that dart upon their quarry by a side glance ; the true falcons being "noble" birds, because they soar to heights unseen, and drop from a perpendicular like a thunderbolt on a selected victim. It was the European counterpart of our duck hawk that furnished royal sport in the Middle Ages.

American sportsmen best know how unerring is the marksmanship of this marauder. The teal, one of the swiftest travellers on wings, will be whistling its way above the sloughs, when, quicker than thought, its throat is seized by an unseen, unsuspected foe dropped from the clouds. It is choked to death even while both birds are falling to the ground ; and in less time than its takes to tell, the "noble" falcon will have torn the feathers from the duck's warm breast, and begun a bloody orgy. Only the fortunate duck attacked above water, into which it may plunge and swim below the surface, stands a reasonable chance of escape. Geese and the larger fowls may be stunned by the blow as the falcon falls upon them ; but not until the assassin, after repeated onslaughts, finally strangles its prey, does the plucky bird cease its heroic fight for life. Little birds are eaten entire, but the entrails of larger ones remain untouched. Following the immense flocks of water-fowl in their migrations, the falcon makes sad havoc among them. It is amazing how large a bird the villain can bear away with ease. Pigeons, Bob Whites, grouse, meadow larks, hares, and herons are conspicuous victims ; but even the courageous crow becomes a limp coward in the neighborhood of this most audacious, fleet-winged, strong-footed rascal. "No bird is more daring," says Mr. Chapman ; "I have had duck hawks dart down to rob me of wounded snipe lying almost at my feet, nor did my ineffective shots prevent them from returning." Ospreys often band together to wreak their vengeance on the eagle, but apparently the falcon pursues his bloody career unmolested. In his presence every bird quakes.

The nest, built on rocky cliffs or in the hollow limbs of tall trees, contains three or four creamy white or fawn colored eggs irregularly blotched, smeared, and streaked with brown and brownish red.

Kites, Hawks, Eagles, etc.

The Pigeon Hawk (*Falco columbarius*), a much smaller filibuster than the preceding, being a foot or less in length, bears some resemblance to it in habits. Without hesitation it will attack a bird of its own or greater size, strangle it, pluck it, and feast upon its breast. Following in the wake of migrant song birds, it keeps an interested eye on a weak or wounded robin, bobolink, or blackbird, to pounce upon it the instant it straggles behind the flock. In the air and when perching, it so closely resembles the passenger pigeon that it has not infrequently been mistaken and shot for one. The pigeon hawk is an equally rapid flyer, and, of course, far more dashing than that rather spiritless bird. As if to be avenged for the misdirected shots that kill its race instead of the pigeons, the hawk eats them whenever it has an opportunity. Open country and the edges of woods, particularly near water, are its favorite hunting grounds throughout a range extending over the whole of North America. As it nests chiefly north of the United States, and spends its winters south, even touching northern South America and the West Indies, it is as a spring and autumn migrant that we know the pigeon hawk here. Its upper parts vary between slaty blue and brownish gray, with a broken rusty or buff collar; its primaries are barred with white; the under parts are buff or pale fawn color, almost white on the throat; the breast and sides have large oblong brown spots, and the tail has three or four grayish white bars and a white tip. As the bird is far from shy, it is not difficult to get a glimpse at the plumage while it perches on a low branch waiting for its prey to heave in sight.

American Sparrow Hawk

(Falco sparverius)

Called also RUSTY CROWNED FALCON; AMERICAN KESTREL; MOUSE HAWK; KILLY HAWK

Length—10 to 11 inches. Sexes the same size.
Male—Top of head slaty blue, generally with a reddish spot on crown, and several black patches on sides and nape; back rufous, with a few black spots or none; wing coverts ashy blue with or without black spots; tail bright rufous, white

AMERICAN SPARROW HAWK.

tipped, and with a broad black band below it, the outer feathers white with black bars; under parts white or buff, sometimes spotted with black.

Female—Back, wing coverts, and tail rufous with numerous black bars; under parts plentifully streaked with dark brown.

Range—Eastern North America, from Great Slave Lake to northern South America. Nests from northern limits of range to Florida; winters from New Jersey southward.

Season—Summer resident in the northern United States and Canada; March to October; winter or permanent resident south of New Jersey.

Perched on a high dead limb, the crossbar of a telegraph pole, a fence post, or some distended branch—such a point of vantage as a shrike would choose for similar reasons—the beautiful little sparrow hawk eagerly scans the field below for grasshoppers, mice, hair sparrows, and other small quarry to come within range. The instant its prey is sighted, it launches itself into the air, hovers over its victim, then drops like a stone, seizes it in its talons, and flies back to its perch to feast. It is amusing to watch it handle a grasshopper, very much as a squirrel might eat a nut if he had but two legs. Or, becoming dissatisfied with its hunting grounds, it will fly off over the fields gracefully, swiftly, now pausing on quivering wings to reconnoitre, now on again, past the thickets on the outskirts of woods, through the orchard and about the farm, suddenly arresting flight to pounce on its tiny prey. Its flight is not protracted nor soaring. Never so hurried, so swift, or so fierce as other small hawks, it is none the less active, and its charming hovering posture gives its flight a special grace. *Kill-ee-kill-ee-kill-ee* it shrilly calls as it flies above the grass. Every farmer's boy knows the voice of the killy hawk. Less shy of men than others of its tribe, showing the familiarity of a robin toward us, and it is certainly more social than most hawks, for one frequently sees several little hunters on the same acre, especially around the bird roosts in the spring and autumn migrations. The sparrow hawk would be a universal favorite were it not for its rascality in devouring little birds. So long as there is a grasshopper or a meadow mouse to eat, it will let feathered prey alone; but these failing, it is a past master in dropping like a thunderbolt upon the tree sparrows, juncos, thrushes, and other small birds found on the ground in thickets and the borders of woods. But it does not eat the farmer's broilers: the little sharp-shinned and the Cooper's hawk attend to them. However, the average farmer, who confounds the sins of the former with the far slighter offences of the sparrow hawk, shoots the bird that destroys more enemies to his prosperity than he could guess. Of the three hundred and twenty stomachs of the sparrow hawk examined by Mr. Fisher for the Department of Agriculture, two hundred and fifteen contained grasshoppers or other large insects, eighty-nine contained mice, and not one contained poultry.

Unlike other birds of prey, the sparrow hawk builds no nest, but lays in the hollows of trees, crevices of rocks, or even about outbuildings, on a farm; but a deserted woodpecker's hole is its ideal home. Although this bird arrives from the south in March, it does not nest until May, when from three to seven cream or fawn-colored eggs, finely and evenly marked with reddish brown, are carefully tended by both the mates that remain lovers for life.

American Osprey
(Pandion haliaëtus carolinensis)
Called also : FISH HAWK

Length—Male 2 feet, or a trifle less; female larger.
Male and Female—Upper parts dusky brown, the feathers edged with white as a bird grows old, head and nape varied with white and a dark stripe on side of head; under parts white, the breast of male sometimes slightly, that of female always, spotted with grayish brown; tail with six or eight obscure dark bars. Bill blackish and with long hook; iris red or yellow; long, powerful feet, grayish blue.
Range—North America from Hudson Bay and Alaska to northern South America and the West Indies; nesting throughout its North American range.
Season—Summer resident, March to October, except in southern part of range.

Is there a more exhilarating sight in the bird kingdom than the plunge of the osprey ? From the height where it has been circling and coursing above the water, it will quickly check itself and hover for an instant at sight of a fish swimming near the surface; then, closing its great wings, it darts like a streak of feathered lightning, and with unerring aim strikes the water with a loud splash. Perhaps it will disappear below for a second before it rises, scattering spray about it in its struggles to clear the surface, and fly upward with its prey grasped in its powerful talons. The fish is never carried tail end foremost; if caught so, the hawk has been seen turning it about in mid air. Small fry are usually eaten a-wing; larger game are borne off to a perch, to be devoured at leisure; and it is said that when an osprey strikes its talons through the flesh of a fish too heavy to be lifted from the water, the prey turns captor and drowns his tormentor, whose claws reaching his vitals soon end his life, when bird and fish, locked in a death grasp, are washed ashore. The osprey rarely touches fish of value for the table; catfish, suckers, and such prey as no one grudges it, form its staple food. It also eats with relish dead fish lying on the beach.

The bald eagle, perched at a high point of vantage, takes instant note of the successful fisher, and with a majestic swoop arrives before the osprey has a chance to devour its prey. Now a desperate chase begins if the intimidated bird has not already relaxed its grasp of the prize; and pursuing the hawk higher and higher, the eagle relentlessly torments it until it is glad to drop the fish for the pirate to seize and bear away, leaving it temporary peace. Again the industrious osprey secures a glistening, wriggling victim; again the eagle pursues his unwilling purveyor. After unmerciful persecution, a number of fish hawks will band together and drive away the robber.

Birds of this order show strong affection for their life-long mates and the young, and for an old nest that is often a true home at all seasons, and to which they return year after year if unmolested, simply repairing damages inflicted by winter storms. The osprey also shows a marked preference for a certain perch to which it carries its prey, and there it will sit sometimes for hours at a time. The ground below is heavily strewn with bones, scales, and other indigestible parts of fish. An immense accumulation of sticks, rushes, weed stalks, shredded bark, salt hay, odds and ends gathered among the rubbish of seaside cottages, feathers, and mud make old nests, with their annual additions, bulky, conspicuous affairs in the tree tops. New nests are often rather small, considering the size of the bird. Both mates incu-

Kites, Hawks, Eagles, etc.

bate the eggs, which are from two to four, extremely variable in size and coloration, sometimes plain dull white, sometimes almost wholly chocolate brown, but normally buff, heavily marked with chocolate, especially around the larger end. Colonies of nesters are frequently reported along our coasts, and instances of a pair of grackles utilizing a corner of the osprey's ample cradle for theirs are not rare. In four weeks or less after their eggs are laid, the fish hawks are kept busy shredding food for their downy, helpless young.

OSPREY.
½ Life size.

BARN OWLS
(Family Strigidæ)

American Barn Owl
(Strix pratincola)

Called also: MONKEY FACED OWL

Length—15 to 18 inches: female the larger.

Male and Female—Upper parts mottled gray and buff finely speckled with black and white; heart-shaped facial disks and under parts whitish or buff, the latter with small round black spots; tail white or buff, mottled with black, and sometimes with three or four narrow black bars like the wings; eyes small, black; no horns; long, feathered legs; long, pointed wings reaching beyond tail.

Range—United States, rarely reaching Canada, south to Mexico, nesting from New York state southward.

Season—Permanent resident, except at northern limit of range.

The American counterpart of "wise Minerva's only fowl," known best by its startling scream, keeps its odd, triangular face, its speckled and mottled downy feathers, and its body, that looks more slender than it really is, owing to its long wings, well concealed by day; and so silently does it move about at night that only in the moonlight can one hope for a passing glimpse as the barn owl sails about on wide-spread, tapering wings, and with a hawk-like movement, from tree to tree. "The face looks like that of a toothless, hooked-nosed old woman, shrouded in a closely fitting hood," says Mrs. Wright, "and has a half-simple, half-sly expression that gives it a mysterious air." Periodically a very old hoax is played on a credulous public by some newspaper reporter who declares that in such a town, by such a man, a curious creature has just been caught, half-bird and half-monkey!

By day, all owls look sleepy and sad; but at dusk, when rats and mice creep timidly forth, the barn owl, now thoroughly awake, sallies from its hole and does greater execution before morning than all the traps in town. Shrews, bats, frogs, grasshoppers, and beetles enlarge its bill of fare. A pair of these mousers that had their nest in an old apple tree near a hayrick that concealed the spectator, brought eight mice to their brood in the hollow trunk in less than an hour.

The head of a mouse, the favorite tid-bit, is devoured first; then follows the body, bolted whole if not too large. One foot usually holds the smaller quarry; but a rat must be firmly grasped with both feet, and torn apart before it is bolted. Since owls swallow skin, bones, and all, these indigestible parts are afterward ejected in pellets. Disturb the owls at their orgy, and they click their bills and hiss in the most successful attempt they ever make to be ferocious. They are not quarrelsome even among themselves when feeding, and the smallest songster can safely tease them to a point that would goad a less amiable bird to rashness. A querulous, quavering cry frequently repeated, k-r-r-r-r-r-ik, suggesting the night jar's call, is sometimes more frequently heard than the wild, peevish scream usually associated with this owl.

In spite of civilization's tempting offers, a hollow tree has ever remained the favorite home of the barn owl, that nevertheless deserves its name, for barns and other outbuildings on the farm, steeples, and abandoned dove cots become equally dear to it once they have sheltered a brood. A pair of these owls have nested for years in one of the towers of the Smithsonian Institution; many eggs have been laid directly on roofs of dwellings; some in mining shafts; others in deserted burrows of ground squirrels and other rodents; in fact, all manner of queer sites are chosen. Strictly speaking, the barn owl builds no nest, unless the accumulation of decayed wood, disgorged bones of mice, etc., among which the eggs are dropped, could be honored with such a name. From five to eleven pure dull white eggs, more decidedly pointed than those of most owls, are incubated by both mates, sometimes by both at once, as they sit huddled together through the hours of unwelcome sunshine. They can scarcely multiply too fast. The barn owl does not eat poultry, although it is constantly shot because of an unfounded belief that it does, prevalent among farmers. From an economic standpoint, it would be difficult to name a more valuable bird.

HORNED AND HOOT OWLS
(Family Bubonidæ)

American Long-eared Owl
(Asio wilsonianus)

Called also: CAT OWL

Length—14 to 16 inches; female the larger.

Male and Female—Conspicuous blackish ear tufts bordered by white and buff; upper parts dusky brown, finely mottled with ash and dull orange; facial disk pale reddish brown with darker inner circle and yellow eyes; under parts mixed white and buff, the breast with long brown stripes, the sides and underneath irregularly barred with dusky, dark broken bands on wings and tail; legs and feet completely feathered; bill and claws blackish.

Range—Temperate North America; nesting throughout its range.

Season—Permanent resident.

A strictly nocturnal prowler, unlike its short-eared relative that hunts much by day, the long-eared owl keeps concealed through the hours of sunshine in the woods, the alder swamps, or high, dry, shady undergrowth, giving no hint by sign or sound of its hiding place. Come upon one suddenly, and by pressing its feathers close to its body, erecting its ear tufts, and sitting erect, it doubtless hopes to be overlooked as a part of the weather-beaten tree on which it is perching, since its thick, downy, mottled plumage might readily be mistaken for rough bark; but as it blinks its staring eyes knowingly, it looks amusingly like a mischievous, round-faced joker, half bird, half human. It is not easily frightened away, and is ever peaceably disposed. To look formidable when liberties are taken with it, it may ruffle up its feathers until its circumference is doubled; but nothing happens, unless it be a noiseless gliding off among the trees to another perch.

At nightfall, it flies with almost uncanny softness, skimming along the ground, exploring leafy avenues and grassy meadows and swamps; its wide, staring eyes, immovably fixed in the sockets, scanning the hunting ground, as the head, inclined downward, turns now this way, now that. Shy, skulking mice, pounced upon unawares by the silent prowler, other small mammals, and very rarely a bird, are carried off in the talons, to be devoured at leisure. Like other owls, this species flies slowly and almost uncertainly, but with a buoyancy that gives no suggestion of effort.

About three-fourths of all nests reported are those built by crows and afterward permanently appropriated by the cat owl. It almost never builds for itself; even a squirrel's nest is preferable to one of its own construction. Three to six white eggs require about three weeks of close incubation.

It is chiefly at the nesting season that these usually silent birds lift up their voices. "When at ease and not molested," says Captain Bendire, "the few notes which I have heard them utter are low toned and rather pleasing than otherwise. One of these is a soft-toned *wu-hunk, wu-hunk*, slowly and several times repeated. . . . Another is a low, twittering, whistling note, like *dicky, dicky, dicky*, quite different from anything usually expected from the owl family. In the early spring they hoot somewhat like a screech owl, and may often be heard on a still evening; but their notes are more subdued than those of the latter." The most common cry of the long-eared owl, the one that has given it its popular name, is a prolonged *me-ow-ow-ow*, so like a cat's cry that it would seem folly for a bird that lives chiefly on mice to utter it.

Short-Eared Owl

(Asio accipitrinus)

Called also: MARSH OWL; MEADOW OWL; PRAIRIE OWL

Length—14 to 17 inches; female the larger.

Male and Female—Ear tufts inconspicuous; face disk white, or nearly so, minutely speckled with blackish, and with large black eye patches and yellow eyes; upper parts dusky brown, the feathers margined with yellow; under parts whitish or buff, the breast broadly streaked, never mottled, with brown, and underneath more finely and sparingly streaked, tail barred with buff and dusky bands of equal width. Bill and claws dusky blue black; legs feathered with buff.

Range—Nearly cosmopolitan; throughout North America, and nesting from Virginia northward.

Season—Chiefly a migratory visitor; April, November; also a resident in many sections.

Here is an owl that breaks through several family traditions, for it does not live in woods, neither does it confine its hunting excursions to the dark hours; but, living in the marshes or grassy meadows, it flies abroad much by day, especially in cloudy weather, after two o'clock in the afternoon, as well as at night. Another unconventional trait it has: it makes its nest of hay and sticks on the ground instead of in hollow trees or upper parts of buildings; and one nest that contained six white eggs, discovered in a lonely marsh where the least bittern was the owl's nearest neighbor, was in a tussock quite surrounded by water. The bittern, that misanthropic recluse, springing into the air, was off at once, dangling its legs behind it; whereas the marsh owl, that is not at all shy, simply stared and blinked, with a half human expression of wonder on its face, until the intruder became too impertinent and lifted it off its nest. Even then it did nothing more spiteful than to sharply click its bill as it circled about just overhead. Yet there seems to be a popular impression that this owl is fierce. Even Nuttall has said it will attack a man! In the west the burrows of ground squirrels and rabbits or the hole of a muskrat have been utilized, since none of the owls is overscrupulous about appropriating other creatures' homes, however much attached a pair may become to a spot that has once cradled their brood. "As useless as a last year's nest" can have no meaning to owls. Still another peculiarity of this owl is that it is almost never seen to alight on a tree; the ground is its usual resting place, a stump or knoll a high enough point of vantage. Mice, gophers, and insects of various kinds, which are its food, keep this hunter close to earth; and as it flies low, and does not take to wing until fairly stepped on, it encourages close acquaintance, thereby earning a reputation for being the most abundant species in the United States. Its alleged superiority of numbers may also be accounted for by the fact that during the migrations it is sometimes found in flocks numbering a hundred.

Aside from a quavering, mouse-like squeak, the marsh owl apparently makes no sound. Its flight is positively uncanny in its silence. Like the barn and the long-eared owls, this invaluable ally earns the fullest protection from the farmers.

Barred Owl

(Syrnium nebulosum)

Called also: HOOT OWL; WOOD OWL

Length—18 to 20 inches; female the larger.

Male and Female—Upper parts grayish brown, each feather with two or three white or buff bars; facial disk gray, finely barred or mottled with dusky; eyes bluish black, and bill yellow; under parts white washed with buff; the breast barred; the sides and underneath streaked with dusky; legs and feet feathered to nails; wings and tail barred with brown, no ear tufts.

Range—Eastern United States to Nova Scotia and Manitoba; west to Minnesota, Nebraska, Kansas, and Texas; nesting throughout range.

Season—Permanent resident.

Whoo-whoo-too-whoo-too-o-o, with endless variation, a deep toned, guttural, weird, startling sound, and *haw-haw-hoo-hoo*, like a coarse, mocking laugh, come from the noisy hoot owl between dusk and midnight, rarely at sunrise, more rarely still by day, sometimes from a solitary hooter, sometimes in a duet sung out of time. Every one knows the hoot. One hears it most frequently at the nesting season. Once in a very great while this owl gives a shriek to make one's blood curdle. Many of us have attracted the bird by imitating its notes. Because the voice of the great horned owl, that "tiger among birds," is so like it, the barred owl is credited with its larger kinsman's atrocities and shot. Its own talons are not wholly guiltless of innocent blood, to be sure, since out of one hundred and nine stomachs examined by Dr. Fisher for the Department of Agriculture, five contained young poultry or game, and thirteen other birds; but over one-third contained mice and other small mammals; frogs, fish, lizards, and insects filled the remainder, which goes to prove that, in spite of the average farmer's belief to the contrary, this owl renders him positive service. To see the barred owl is to identify it at once by its smooth, bland, almost human face, its mild blue black eyes, and the absence of horns from its round head.

Woods, waysides, and sheltered farms are the barred owl's hunting grounds; and because it so frequently lodges through the sunny hours in hay lofts and stacks, many people call it the barn owl, a name which should be discouraged by disuse to save the endless confusion arising by the application of the same popular name to two or more different species. True barn owls are not only a distinct species, but constitute a separate family. The unearthly, weird voices of several owls make each one indifferently a "hoot owl" to the average listener.

In February, the barred owl loses his unsocial, hermit-like instinct, and for his mate's society, at least, shows a devoted preference. The pair go about looking for a natural cavity in a tree in dense or swampy woods; but that failing them, they unscrupulously take possession of a hawk's or crow's nest, tenaciously holding it year after year, as all owls do their homes. They rarely build a nest of their own, or take pains to line a cavity or to alter an appropriated tenement unless it should need repairs. Mr. C. L. Rawson has found a set of eggs lying on a solid cake of ice near Norwich, Conn., so early is the nesting done. A camera can take no more amusing picture than a group of owlets perching on a naked limb near their cradle, their downy feathers ruffled by the March wind, a surprised, comical expression on their faces, their bodies closely huddled together to save warmth.

The Great Gray, or Spectral Owl (*Scotiaptex cinereum*), the largest owl in the world, is dusky, mottled with white on its upper parts, and the white under parts are broadly streaked on the breast, and on the sides and underneath irregularly barred and streaked with dusky. It has no ear tufts; its legs and feet are heavily feathered, and both bill and eyes are yellow. It is a very rare visitor from the far north, and as it keeps to dense woods, few bird students have been so fortunate as Dr. Dall, who has caught it in his hands. He declares it is "a stupid bird." No owl that is heavy with sleep while humans are wide awake is credited with great cleverness, and modern ornithologists consider that Minerva made a mistake when she chose the owl to typify that wisdom for which she was revered.

Saw-Whet Owl

(Nyctala acadia)

Called also: ACADIAN OWL

Length—7.50 to 8 inches; smallest of the eastern owls.

Male and Female—Upper parts dark reddish brown, the head streaked, the back and wings spotted with light brown and white; under parts white, heavily streaked with dark rusty brown; tail with three or four broken white bars; facial disk almost white, but blackened around the yellow eyes; legs buff feathered; no ear tufts; bill and claws dark.

Range—North America at large, nesting from the middle states northward, and in western mountainous regions south to Mexico.

Season—Chiefly a winter visitor in middle states; locally a permanent resident.

Birds that prey on skulking mice must needs be night prowlers; but the theory has been seriously advanced that those owls which remain in sleepy seclusion all day, like the little saw-whet owl, are those that have suffered endless persecution from other birds whenever they ventured abroad in the sunlight, which is the reason they choose to hunt when others sleep—an interesting theory, if nothing more. Nocturnal birds are naturally counted rarities, even where they are not. This little owl, by no means uncommon, is a very sound sleeper, and makes no sign of its existence, though one may be passing beneath its perch. So sunk in oblivion is it, and so heavy-eyed with sleep when roused, that many specimens may be taken with the hand. Hence the expression, "As stupid as an owl." Because it is small enough to crowd on a woman's hat, this is a little victim commonly worn, sometimes with wings and tail outspread, or again with only its head, like a Cheshire cat's, appearing in a cloud of trimming. A woman who loudly applauded Dr. Van Dyke's famous epigram, "A bird in the bush is worth two in the hat," at an Audubon Society's meeting held in the Museum of

SAW-WHET OWL.
⅔ Life-size.

Natural History, New York, in the winter of 1898, had the entire plumage of a saw-whet owl spread over her turban! At this same meeting another woman with self-righteous superiority was overhead boasting that she never wore birds, only wings, in her hats!

Saw-whet, saw-whet, the love notes of this owl, most frequently heard in March and April, have a rasping quality like the sound heard in a mill when the file is sharpening the teeth of a saw; not an agreeable noise, perhaps, yet because of the ventriloqual power of the bird's voice, and at the distance we think we hear it, it has a certain fascination.

Dense woodlands, particularly evergreen forests, for it dearly loves a dark retreat, are where the owl passes its days; coming out at night, when its flight, surprisingly like a woodcock's, has deceived others than Dr. Fisher into making a worse than wasted shot. Since it feeds almost exclusively on mice and insects, it is folly to destroy so valuable a bird. Mr. Nelson says that a dozen specimens have been taken in the most frequented streets in the residence portion of Chicago within two years.

The majority of nests recorded have been deserted woodpecker's holes, some in squirrel's excavations, most of them near water, a few in stumps; but very rarely do the saw-whet owls appropriate an open nest, and more rarely still build one of their own.

Screech Owl

(Megascops asio)

Called also: MOTTLED OWL; RED OWL; LITTLE HORNED OWL

Length—8.50 to 9.50 inches.
Male and Female—Brownish red phase: Upper parts rusty red, finely streaked with blackish brown and mottled with light brown; under parts whitish or buff, the feathers centrally streaked with black and with irregular rusty bars. Eyes yellow; legs and feet covered with short feathers; prominent ear tufts. *Gray phase:* Upper parts ashen gray streaked with black and finely mottled with yellow; under parts white, finely streaked and barred irregularly with black, more

or less bordered with rusty. Immature birds have entire plumage regularly barred with rusty, gray, and white.

Range—Eastern North America.
Season—Permanent resident.

Why this little owl should wear such freaky plumage, rusty red one time, mottled gray and black another, without reference to age, sex, or season, is one of the bird mysteries awaiting solution. Frequently birds of the same brood will be wearing different feathers. In the transition from one phase to another, many variations of color and markings appear; but however clothed, we may certainly know the little screech owl by its prominent ear tufts or horns, taken in connection with its small size. Like the little saw-whet owl, which, however, wears no horns, people who live in cities are most familiar with it on women's hats, worn entire or cut up in sections.

A weird, melancholy, whistled tremulo from under our very windows startles us, as the uncanny voices of all owls do, however familiar we may be with the little screecher. Are any superstitions more absurd than those associated with these harmless birds? Because it makes its home so near ours, often in some crevice of them, in fact, in the hollow of a tree in the orchard, or around the barn lofts, this is probably the most familiar owl to the majority of Canadians and Americans. It keeps closely concealed by day, often in a dense evergreen or in its favorite hollow; and except for the persecutions of the blue jay, that takes a mischievous delight in routing it from its nap and driving it abroad for all the saucy birds in the orchard to pursue and peck at, we should never know of its presence. In the early spring especially it lifts up its voice—too doleful a love song to be effective, one would think; yet the screecher's mate apparently considers it entrancing, since she remains mated for life. In the southern and central portions of its range, nesting begins in March; in the New England and northern parts some time between the middle of April and the first of May. A natural cavity in a hollow tree, or an abandoned woodpecker's hole are favorite nooks, and boxes nailed up under the dark eaves of outbuildings on the farm or in dense evergreen trees where light cannot strike the owl's sensitive eyes, have been promptly appropriated in many instances.

It is generally known that all owls go through some strange

SCREECH OWL.

performances to woo their mates, but few have been so fortunate as Mr Lynds Jones, who watched a pair of screech owls mating. "The female was perched in a dark, leafy tree," he says, "apparently oblivious of the presence of her mate, who made frantic efforts, through a series of bowings, wing-raisings, and snappings, to attract her attention. Those antics were continued for some time, varied by hops from branch to branch near her, accompanied by that forlorn, almost despairing wink peculiar to this bird. Once or twice I thought I detected sounds of inward groanings, as he, beside himself with his unsuccessful approaches, sat in utter dejection. At last his mistress lowered her haughty head."

When hunting, the owl moves like a shadow, so silently does it pass in the darkness. Insects, cut worms, and mice are what it is ever seeking; but sharp hunger in winter has sometimes led it into butchery of little birds. Of two hundred and fifty-five stomachs of screech owls examined by Dr. Fisher for the Department of Agriculture, one hundred contained insects; ninety-one, mice; thirty-eight, birds; eleven, other mammals than mice; nine, crawfish; seven, miscellaneous food; five, spiders; four, batrachians; two, lizards; two, scorpions; two, earth worms; one, poultry; one, fish; and forty-three were empty. Why in the name of all that is economic and humane, should this valuable ally of the farmer be so persistently shot?

Great Horned Owl

(Bubo virginianus)

Called also: HOOT OWL; CAT OWL

Length—Male 19 to 23 inches; female 21 to 26 inches.
Male and Female—Long ear tufts; upper parts variegated brown, tawny, pale buff, and white; facial disk buff; eyes yellow; throat white; under parts buff or whitish, finely barred with black; legs and feet feathered.
Range—Eastern North America, west to the Mississippi, and from Labrador to Costa Rica.
Season—Permanent resident.

The lord high executioner of the owl tribe, remaining a permanent resident, except at the extreme northern limit of his range, does more damage than all other species put together. Although actually shorter than the great gray and snowy owls, his ponderous body gives him more impressive size and power, earned through constant exercise of savage instincts. No one ever finds this hunter in poor condition; diligent and overpowering in the chase, he feasts where others starve, bringing down upon the innocent heads of several members of his tribe the punishment of sins of his commission by undiscriminating farmers. Only the sharp-shinned, the Cooper's hawk, and the goshawk among the birds of prey can show a bloodier record.

By day this "tiger among birds" keeps concealed in the woods, particularly among evergreens near water, in cloudy weather sometimes sallying forth for food, but generally not until dusk; then, with uncanny silence and hawklike swiftness of flight, he begins his nefarious work. Chickens, ducks, geese, turkeys, and pigeons on the farm will be decapitated if too large to eat entire, for the brains of victims are the tid-bits this executioner delights in. Dr. Hart Merriam tells of one of these owls that took the heads off three turkeys and several chickens in a single night, leaving their bodies uninjured and fit for the table. Coops and dove cots are boldly entered; entire coveys of Bob Whites destroyed; grouse, woodcock, water-fowl, and snipe know no more relentless enemy; song birds do not escape the stealthy murderer that picks them from the perch as they sleep; and all the rats, mice, squirrels, rabbits, and other mammals eaten cannot offset the valuable birds destroyed.

A piercing scream as of a woman being strangled, the most blood-curdling sound heard in the woods, a rare but all too frequent sound, is a fitting vocal expression of a character so unholy.

> "Silence, ye wolves! while Ralph to Cynthia howls
> And makes night hideous;—answer him, ye owls."

A deep-toned *to-whoo-hoo-hoo, to-whoo-whoo*, as of a hound baying in the distance, louder than the barred owl's hoot, and the syllables all on one note, is the sound so familiar as to scarcely need description. Like all owls, this one seems particularly attracted by the camp-fire, and every sportsman knows how dismally it punctuates the silence of the woods.

Unsocial, solitary, except at the nesting season, unapproach-

GREAT HORNED OWL.
½ Life-size.

able by men, unlike several of his kin that may be taken in the hand when sleeping, the great horned owl gives one little opportunity for close acquaintance in his wild state, and because he is an irreclaimable savage in captivity few keep him caged. With eyes closed so as to leave only a crack to peek through, one might think he did not see the intruder; but go to right or left, and the head turns around so far to note every step that it must seemingly drop off. All owls have eyes immovably fixed in the sockets, which is the reason they must almost wring their necks when they attempt to look around. The large, yellow iris of this owl is capable of extraordinary contraction; but before you can fairly see its interesting operations, the great bird spreads his wings and moves through the trees with the silence of a shadow of a passing cloud.

In February, when the nesting season begins—for it is supposed this owl breaks the family rule and does not remain mated for life—he singles out some sweetheart, always larger and more formidable than himself, and undertakes the difficult task of wooing her. At first remaining an indifferent spectator of his ludicrous leaps and bounds on the earth and from tree to tree, his eccentric evolutions in the air, and the rapid snappings of his bill, she finally relents and goes house hunting with her attentive escort. Hollow trees with entrance large enough for their bodies are scarce; and when not to be found, an old crow's, hawk's, or squirrel's nest is utilized. Two or three dull white eggs are laid so early in the year that ice not infrequently makes them sterile, in which case they simply contribute to the accumulation of rubbish at the bottom of the nest, on which a new set is laid. Oftentimes the nesting is over with in time to allow the rightful owner of the cradle, or one of the larger hawks, to use it. A careful observer tells of finding in a nest containing two young owls "a mouse, a young muskrat, two eels, four bullheads, a woodcock, four ruffed grouse, one rabbit, and eleven rats. The food taken out of the nest weighed eighteen pounds. A curious fact connected with these captives was that the heads were eaten off, the bodies being untouched."

Snowy Owl

(Nyctea nyctea)

Length—About 2 feet.
Male and Female—White, more or less barred or spotted with dusky; some specimens almost entirely white, the female usually the more heavily barred; the face, throat, and feet being in all birds the whitest parts; legs and feet thickly feathered; iris yellow; bill and claws black; no ear tufts.
Range—"Northern portions of northern hemisphere. In North America breeding wholly north of the United States; in winter migrating south to the middle states, straggling to South Carolina, Texas, California, and Bermuda." A. O. U.
Season—Winter visitor.

No Arctic explorer has yet penetrated too far north to find the Snowy Owl. Private Long of the Greely expedition, who raised six of these owlets, released them only because food became scarce enough for men during the second winter of hardship, much less for such greedy pets. "They had inordinate appetites," says the commander, "and from the time they were caught, as young owlets, swallowed anything given to them. I remember one bolting whole a sandpiper about half his own size. Over a hundred and fifty skuas (robber gulls) were killed and fed to these owls. It was interesting to note that, although they had never used their wings, the owls flew well." In another volume, General Greely describes the snowy owl's egg as "somewhat larger than, though closely resembling, the white egg of a hen. Sergeant Israel found it very palatable. The male bird showed signs of fight when the egg was taken, while the female looked on from about one hundred yards. The first owl observed was on April 29th, since then one or more have been frequently seen. The nest is a mere hole hollowed out on the summit of a commanding knoll, and furnished with a few scattered feathers, grass, etc."

The lemming, ptarmigan, ducks, and other water-fowl are the snowy owl's main dependence. It is an expert fisher, too, and borne up by the seaweed, it patiently waits for finny prey to swim among the rocks, when, quicker than thought, they are captured. The Arctic hare, though double its own weight, is pounced upon and immediately lifted off its feet lest it use its strong hind legs to escape from its captor. The ptarmigan, on the contrary, is "crushed to earth," but, unlike truth, it may not rise again; for the owl, spreading its great wings to render those of its victim useless, soon ends its struggles. A bird must be a swift flyer, indeed, that can overtake a duck in the air or a hare afoot. The former it strikes down after the manner of the goshawk.

But when food begins to fail at the far north, this hardy owl that is able to endure the most intense cold—since all do not migrate by any means—leaves the moss and lichen-covered tundras, and, joining a band of travellers bound southward, appears in the United States sometimes in considerable numbers, especially in the Atlantic states. A northeasterly storm drives many migrants ashore. Some morning when trees and earth are covered with a snowy mantle that the high winds toss and blow, you may see a wraith, a ghostly apparition, gleaming at you with fiery eyes from the evergreen. Here is a miracle of nature: the snow is alive! Winter incarnate sits before you. Juncos and snow buntings whirl about among the snowflakes in scattered flocks; and the wraith, as silently as any spectre, its downy wings outspread, floats off from its high point of vantage in easy circles, then suddenly swooping, seizes a hapless bird in its talons. In boldness and grace of flight it is far more like a hawk than an owl; and, moreover, it is a diurnal bird of prey, comparatively little of its hunting for mice and other food being done at night.

The Hawk Owl (*Surnia ulula caparoch*), another northern species that occasionally visits the United States border or beyond, as far as Pennsylvania, in winter, is of medium size (fifteen inches) and without horns. Its upper parts are ashen brown, the head and nape spotted with white, and the back and some wing feathers barred with white; the remarkably long tail has rounded white bars. A dusky spot and below it a white one mark the middle throat, while the sides of the neck and upper breast are streaked with dusky, and the rest of the under parts are barred with dusky and white. The legs and feet are feathered, and the wings when folded fall far short of the end of the tail. Like a hawk in habits, as in appearance, it is nevertheless an owl, though doubtless the connecting link between the families. While it hunts its prey chiefly by daylight, it flies with the ghostlike silence characteristic of its clan, yet with a swiftness, boldness, and dash which would lead the uninitiated to suppose it a true hawk. "When the hunters are shooting grouse," says Dr. Richardson, "this bird is occasionally attracted by the report of a gun, and is often bold enough, on a bird being killed, to pounce down upon it, though it may be unable from its size to carry it off. It is also known to hover around the fires made by the natives at night." Its note is said to be a shrill cry, which is generally uttered while the bird is on the wing.

Burrowing Owl

(Speotyto cunicularia hypogæa)

Called also: PRAIRIE OWL

Length—9.50 inches.

Male and Female—Upper parts dull grayish brown spotted with white; wing quills and tail marked with cross rows of spots, sometimes confluent into bars; eyebrow, chin, and throat white, the two latter divided by a dark brown collar; under parts white or pale buff barred with brown spots. Tail short; head smooth, with no ear tufts; facial disk incomplete; legs extremely long and slim, with few or no feathers.

Range—Western United States, from the Pacific coast east through the Great Plains, north into Canada, south to Central America; accidental in New York and Massachusetts.

Season—Permanent resident; or winter visitor at the southern end of its range.

Amusing fictions of this tiny owl living in brotherly love with prairie dogs and rattlesnakes had a serious explosion when Dr. Coues published his "Birds of the Northwest;" but fictions die lingering deaths, and one still reads of "happy families," with the prairie owl cutting a conspicuous figure—groups that Barnum would have certainly secured for the Greatest Show on Earth had they ever existed. "From an extended acquaintance with the habits of the burrowing owl," says Captain Bendire, writing for the government, "I can most positively assert, from personal experience and investigation, that there is no foundation based on actual facts for these stories, and that no such happy

SNOWY OWL.

families exist in reality. I am fully convinced that the burrowing owl, small as it is, is more than a match for the average prairie dog, and the rattlesnake as well; it is by no means the peaceful and spiritless bird that it is generally believed to be, and it subsists, to some extent at least, on young dogs, if not also on the old ones."

Enlarging the deserted burrows of numerous small quadrupeds, especially of ground squirrels, prairie dogs, and badgers, but never living with them, the burrowing owl begins at the far end of the tunnel to loose the earth and send it backward with vigorous kicks until all is clear. Now dry horse or cow dung is carried to the burrow, broken up in little pieces, and scattered over the nesting chamber, which may be eight or even ten feet from the entrance. In California, dry grass, feathers, weed stalks, and such material may serve as a carpet; but however lined, a prairie owl's nest is sure to abound in fleas. These sometimes speckle the eggs until no one who had not washed them would suspect they were normally a clear, glossy white. As many as eleven eggs have been taken from a chamber where they were found arranged in the form of a horse-shoe—which is usual—or piled in a double layer when the set is exceedingly large; and so skilfully do both mates cover the eggs, that only very rarely does one become addled. The devoted mates remain paired for life.

Except when they have nursery duties to house them—and these are usually ended by June—one may expect to see the funny, top-heavy little owls around the entrance to their homes at any hour of the day, for they are diurnal in habits, and not night prowlers only, although their activities greatly increase after sundown. Bowing toward you as you approach—your entertainer is not shy—a little gnome-like creature nearly twists its head off its neck in its attempt to follow your movements with its immovable eyes. Approach too near, and off it flies, chattering, but not often further than a neighbor's burrow. Small colonies of these owls live in perfect harmony. Doubtless it is the chattering of a disturbed bird, which does sound something like the rattle of a snake, that has given rise to the yarn about rattlesnakes living in owl burrows. Quite a concert of mellow *coo-c-o-o-o-os*, the love note of the species, is kept up by several pairs, sometimes for an hour or more. These notes are uttered only when the birds are at rest and happy. *Zip, zip,* they shrilly cry when alarmed; and besides these sounds, they sharply and rapidly click their bills when excited and enraged, just as all owls do.

A great many colonies are purposely located on the outskirts of prairie dog towns for obvious reasons. The amount of food required by a family of eight or ten hungry owls, any one of which eats more than its weight in a day, is enormous; and after sundown, one sees the busy hunters on the chase, now poised in mid air, like the sparrow hawk, above their prey; now swooping downward on swift, noiseless wings to grasp it in their talons and bear it away. A few well directed blows with the beak, that break the vertebræ of the neck, quiet the struggling victim forever. It is amazing how large the prey is that one of these bold little owls will quickly overcome. The brains are the favorite tid-bits, and often all other parts remain untouched. "Noxious vermin," that is, mice, squirrels, gophers, and such humble quarry, are the prairie owl's staple food, which lays the western farmer under special obligation to see that his rapacious ally receives the fullest protection.

INDEX

The figures in black-faced type indicate the page upon which the life history of the bird is given.

Arrie, **21**.
Auk, Great, 24.
   Little, **25**.
   Razor-billed, 7, **23**.
Auks, The, 5, 6, 18.
Avocet, American, 192, **198**, 209.
Avocets, The, **192**, 198.

Baldpate, 84, **100**.
Beetle-head, 194, **237**.
Black-head, **118**.
   Ring-billed, **120**.
   Ring-necked, **120**.
Bird, Beach, 219.
   Bog, 201.
   Brant, 249.
   Brown, 212.
   Devil's, 68.
   Doe (Eskimo Curlew), 193, **236**.
   Doe (Marbled Godwit), **221**.
   Dough, 236.
   Dun, 131.
   Egg, 21.
   Frost, 239.
   Hay, 212.
   Heart, 249.
   Ice, 25.
   Qua, 166.
   Shad, 204.
   Tortoise-shell, **159**.
Birds, Columbine, **293**.
   Diving, 5–26.
   Gallinaceous Game, **259**.
   Marsh, **171**.
   of Prey, **301**.
   Shore, **191**.
   Surf, 191, **194**, 249.
Bittern, American, 152, **157**.
   Booming, 157.
   Least, 152, **159**, 339.
   Little, **159**.
Bitterns, The, **152**, 157.
Black Breast, **217**.

Bluebill, **118**.
   Little, **120**.
   Marsh, **120**.
Blue Peter, **186**.
Blue Stocking, **198**.
Bob White, 260, **261**, 273.
   Florida, 266.
   Texan, 266.
Bob Whites, The, **259**, 261.
Bog-sucker, **201**.
Booby, **130**.
Brant, 86, **140**.
   Black, 86, **142**.
   Black (White-fronted Goose), **134**.
   Gray, **134**.
   Prairie, **134**.
   White, **136**.
Brent, **140**.
Broadbill (Greater Scaup Duck), **118**.
   (Shoveler), 85, **107**.
   Bastard, **120**.
   Creek, 85, **120**.
Brown Back, **206**.
Brown Jack, **206**.
Bufflehead, 85, 91, **124**.
Bull, Bog, **157**.
Bull-head (Black-breasted Plover), **237**.
Bull-neck, **116**.
Burgomaster, 30, **36**.
Butter-ball, 85, **124**.
Butter-box, **124**.
Buzzard, Red-shouldered, **319**.
   Red-tailed, **317**.
   Swainson's, **321**.
   Turkey, 155, **304**.

Calico-back, 194, **249**.
Chalkline, **164**.
Chicken, Mother Carey's, **68**.
   Prairie (Pinnated Grouse), 260, **278**.
   Prairie (Prairie Sharp-tailed Grouse), **281**.
Chuckle-head, **164**.

Index

Cobb, **37**.
Cock, Black spotted Heath, **270**.
   May, **237**.
   of the Plains, 260, **285**.
   Sage, **285**.
Cockawee, **125**.
Coffin Carrier, **37**.
Coot, 162.
   American, 173, **186**.
   Black, 85, **130**.
   Broad-billed, **130**.
   Butter-billed, **130**.
   Cinerous, **186**.
   Sea, **130**.
   White-billed, **186**.
   White-winged, **131**.
Coots, The, 171, **172**, 177.
Cormorant, Double-crested, 76, 77.
Cormorants, The, **75**.
Coulterneb, **18**.
Crake (Little Black Rail), **183**.
   (Sora), **180**.
Crane, Blue, 152, **161**.
   Brown, 172, **174**.
   Sandhill, 138, 172, **174**.
   Sandhill (Great Blue Heron), **161**.
   White, **176**.
   Whooping, 172, **176**.
Cranes, The, **171**, 174.
Crooked-bill, **235**.
Crow, Carrion, **307**.
Cu-Cu, Large, **223**.
   Little, **224**.
Curlew, Buzzard, **234**.
   Eskimo, 193, **236**.
   Hudsonian, **235**.
   Jack, 193, **235**.
   Little, **236**.
   Long-billed, 193, **234**.
   Red, **221**.
   Short-billed (Eskimo Curlew), **236**.
   Short-billed (Whimbrel), **235**.
   Spanish (Long-billed Curlew), **234**.
   Spanish (White Ibis), 151, **153**.
   Straight-billed, **221**.
Cut-Water, **59**.

Dabchick, 6, **11**.
Darter, Big Blue, **314**.
   Little Blue, **312**.
Diedapper, **11**.
Dipchick, **11**.
Dipper (Horned Grebe), **9**.
   Little, **124**.
   (Pied-billed Grebe), **11**.
Diver, Dun, **87**.
   Great Northern, **14**, 16.
   Hell (Horned Grebe), **9**.
   Hell (Pied-billed Grebe), **11**.
   Pigeon, **25**.
   Red-throated, **16**.

Divers, Lobe-footed, **8**.
Dove, Carolina, **296**.
   Greenland, **25**.
   Mourning, 293, **296**.
   Sea, 7, **25**.
   Turtle, **296**.
Doves, The, **293**, 294.
Dovekie, 7, **25**.
Dowitchee, **206**.
Dewitcher, 193, **206**.
   Long-billed, 193, **208**.
   Western, **208**.
Duck, Acorn, **111**.
   Barrow's Golden-eye, **123**.
   Black, 84, **97**.
   Bridal, **111**.
   Buffalo-headed, **124**.
   Canvasback, 85, 101, 114, **116**.
   Crow, **186**.
   Domestic, **93**.
   Dusky, 84, **97**.
   Dusky Mallard, **97**.
   Eider, 85, **128**.
   Fishing, **87**.
   Golden-eye, 85, 91, **121**.
   Gray, 84, **98**.
   Greater Scaup, 85, **118**.
   Harlequin, 85, **127**.
   Lesser Scaup, 85, **120**.
   Long-tailed, **125**.
   Mallard, 84, **93**.
   Old Squaw, 85, **125**.
   Pintail, 84, **109**.
   Raft, **118**.
   Red-headed, 85, **114**.
   Ring-necked, 85, **120**.
   Ruddy, 65, **131**.
   Scaup, **118**.
   Sea, **128**.
   Spine-tailed, **131**.
   Spirit (Bufflehead), **124**.
   Spirit (Horned Grebe), **9**.
   Sprigtail, **109**.
   Summer, **111**.
   Teal, Blue-winged, 84, **105**.
   Cinnamon, 105.
   Green-winged, 84, **103**.
   Salt Water, **131**.
   Summer, **105**.
   White-faced, **105**.
   Tree, **111**.
   Velvet, **131**.
   Wild, **93**.
   Winter, **109**.
   Wood, 84, 91, **111**.
Ducks, Fishing, **83**, 87.
   River and Pond, 83, **84**, 93, 187.
   Sea and Bay, 83, **84**, 114.
Dunlin, **217**.

Eagle, American, **326**.

Index

Eagle, Bald, 302, **326**, 333.
    Bald Sea, **326**.
    Black, **326**.
    European Golden, **325**.
    Golden, 302, **324**.
    Gray, **326**.
    Mountain, **324**.
    Ring-tailed, **324**.
    War, **324**.
    Washington, **326**.
    White-headed, **326**.
Eagles, The, **301**.
Egret, Blue, 152, **163**.
    White, 152, 156, **164**.
Egrets, The, **151**.

Falcon, Mountain, **328**.
    Peregrine, **328**.
    Rock, **328**.
    Rusty-crowned, **330**.
    Wandering, **328**.
    Winter, **319**.
Falconry, **329**.
Fowl, Flocking, **118**.
    Jungle, **260**.
Flamingoes, **151**.
Fly-up-the-Creek (Least Bittern), **159**.
    (Little Green Heron), **164**.
Fute, **236**.

Gadwall, 84, **98**.
Gallinule, Common, 173, **184**.
    Florida, 173, **184**.
    Purple, 173, **186**.
Gallinules, The, **171**, **172**, 177.
Garbill, **89**.
Garrot, **121**.
Geese, The, 85, **134**.
Godwit, Great Marbled, **221**.
    Hudsonian, **222**.
    Marbled, 193, **221**.
    Red-breasted, **222**.
    Rose-breasted, **222**.
    White-rumped, **222**.
Goosander, 83, **87**.
Goose, Barnacle, **140**.
    Blue-winged, **136**.
    Brant, **140**.
    Canada, 86, **137**.
    Diving, **87**.
    Gray, **137**.
    Laughing, **134**.
    Lesser Snow, 86, **137**.
    Sea, **196**.
    Snow, 86, **136**.
    White-fronted, 86, **134**.
    Wild, 86, **137**.
Goshawk, American, 302, **315**.
Gray Back, **209**.
Great Head, **121**.
Grebe, Carolina, **11**.

Grebe, Dusky, **9**.
    European, Red-necked, **8**.
    Holbœll's, 6, **8**.
    Horned, 6, **9**.
    Little, **11**.
    Pied-billed, 6, **11**.
    Red-necked, **8**.
Grebes, The, **5**, 9, 10, 13.
Greenback, **239**.
Green Head, **93**.
Grouse, Blue, 260, **267**.
    Canada, 260, **270**.
    Canadian Ruffed, 260, **277**.
    Columbian Sharp-tailed, 260, 282, **284**.
    Dusky, 260, **267**.
    Family, **259**, 261.
    Fool, **267**.
    Gray, **267**.
    Gray Ruffed, 260, **277**.
    Mountain, **267**.
    Oregon, 260, **277**.
    Pine, **267**.
    Pinnated, 260, **278**.
    Pin-tailed, **281**.
    Prairie Sharp-tailed, 260, **281**.
    Red Ruffed, 260, **277**.
    Richardson's, **267**.
    Ruffed, 260, 261, **272**.
    Sage, 260, **285**.
    Sharp-tailed, 260, **282**, **284**.
    Sooty, 267, **268**.
    Spotted, **270**.
    Spruce, **270**.
    Willow, **281**.
    Wood, **270**.
Guillemot, **22**.
    Black, 7, **20**.
    Brünnich's, **21**.
    Foolish, 21, **23**.
Gull, Black-headed, **42**.
    Bonaparte's, 30, **44**.
    Burgomaster, 30, **36**.
    Flood, **251**.
    Glaucous, 30, **36**.
    Great Black-backed, 30, **37**.
    Herring, 30, **39**, 42.
    Ice, **36**.
    Iceland, **37**.
    Kittiwake, 30, 32, **35**.
    Laughing, 30, **42**.
    Mackerel, **49**.
    Ring-billed, 30, **41**.
    Risible, **42**.
    Rosy, **44**.
    Summer, **49**.
    Winter, **39**.
Gulls, The, 29, **30**, 35, 43, 50.

Hagdon, **66**.
Hairy Head, **91**.

Index

Harrier, 302, 311.
    Marsh, **311**.
Hawk, 162.
    Blue, **311**.
    Blue Hen, **315**.
    Broad-winged, 302, **322**.
    Chicken (Cooper's), **314**.
    Chicken (Red-shouldered), **319**.
    Chicken (Red-tailed), **317**.
    Cooper's, 302, 313, **314**.
    Duck, 302, **328**.
    Fish, 302, **332**.
    Great Footed, **328**.
    Hare-footed, **323**.
    Hen (Red-shouldered), **319**.
    Hen (Red-tailed), **317**.
    Killy, **330**.
    Marsh, 302, **311**.
    Mouse (Marsh Hawk), **311**.
    Mouse (Sparrow Hawk), **330**.
    Partridge, **315**.
    Pigeon, 302, **330**.
    Pigeon (Sharp-shinned Hawk), **312**.
    Red, **317**.
    Red-shouldered, 302, 317, **319**.
    Red-tailed, 302, **317**.
    Rough-legged, 302, **323**.
    Sharp-shinned, 302, **312**.
    Snake, **309**.
    Sparrow, 302, **330**.
    Swainson's, 302, **321**.
    Western Red-tailed, **319**.
    Winter, **319**.
Hawks, The, **301**.
Hawking, **329**.
Hen, Fresh-water Marsh, **178**.
    Fresh-water Mud, **179**.
    Indian, **157**.
    Marsh, (American Bittern), **152**, **157**.
    Marsh (Clapper Rail), **177**.
    Marsh (King Rail), **173**.
    Meadow, **186**.
    Moor, **186**.
    Mud (American Coot), 175, **186**.
    Mud (Clapper Rail), **177**.
    Mud (Sora Rail), **180**.
    Night, **147**.
    Pine, **267**.
    Prairie, **278**.
    Red-billed Mud, **184**.
    Salt-water Meadow, **177**.
    Water, **184**.
Hern, **162**.
Heron, Black-crowned, Night, 152, **166**.
    Freckled, **137**.
    Great Blue, 152, 156, **161**, 174.
    Green, 152, **164**.
    Little Blue, 152, **163**.
    Snowy, 152, **164**.
    Tribe, 151, **171**.

Herons, The, **152**, 157.
Holopode, Lobe-footed, **196**.
Honker, **137**.

Ibis, White, 151, **153**.
    Wood, 152, **155**.
Ibises, The, **151**, 153.
    Wood, **151**, 155.

Jaeger, Arctic, **32**.
    Long-tailed, 29, **34**.
    Parasitic, 29, **32**.
    Pomarine, 29, **34**.
    Pomatorhine, **34**.
    Richardson's, **32**.
Jaegers, The, **29**.

Kestrel, American, **330**.
Kildee, **242**.
Kildeer, 194, **242**.
Kite, Fork-tailed, **309**.
    Swallow-tailed, 302, **309**.
Kites, The, **301**.
Knot, 193, **209**.
Kricker, **212**.

Lark, Sand, **232**.
Lawyer, **199**.
Leadback, **217**.
Longshanks, **199**.
Look-up, **157**.
Loom, **14**.
Loon, **14**.
    Black-throated, 6, **16**.
    Common, 6, **14**.
    Red-throated, 6, **16**.
    Sprat, **16**.
Loons, The, 5, **6**, 14.
Lords and Ladies, **127**.

Man-of-War, **32**.
Marlin, **221**.
    Brown, 193, **221**.
    Ring-tailed, **222**.
Merganser, American, 83, 87.
    Hooded, 83, **91**.
    Red-breasted, **83**, **89**.
Mergansers, The, **83**, 87.
Murre, Brünnich's, 7, **21**.
    Californian, 7, **23**.
    Common, 7, **22**.
Murres, The, 5, **6**, 18.

Night Peck, **201**.

Old Billy, **125**.
Old Injun, **125**.
Old Molly, **125**.
Old Squaw, **125**.
Old Wife, **125**.
Ortolan (Sora rail), **180**.

Index

Ortolan (Reed Bird), 181.
Osprey, American, 302, 303, **332**.
Owl, Acadian, 303, **342**.
    Barn, 302, **335**.
    Barred, 303, **340**.
    Burrowing, 303, **350**.
    Cat (Great Horned Owl), **345**.
    Cat (Long-eared Owl), **337**.
    Great Gray, **341**.
    Great Horned, 303, **345**.
    Hawk, 303, **349**.
    Hoot (Barred Owl), 303, **340**.
    Hoot (Great Horned Owl), **345**.
    Little Horned, **343**.
    Long-eared, 303, **337**.
    Marsh, **338**.
    Meadow, **338**.
    Monkey-faced, 302, **335**.
    Mottled, **343**.
    Prairie (Burrowing Owl), **350**.
    Prairie (Short-eared Owl), **338**.
    Red, **343**.
    Saw-whet, 303, **342**.
    Screech, 303, **343**.
    Short-eared, 303, **338**.
    Snowy, 303, **348**.
    Spectral, **341**.
    Wood, **340**.
Owls, Barn, The, **302**, **335**.
    Horned and Hoot, The, **303**, **337**, **344**, **347**.
Oyster-Catcher, American, **195**, **251**.
    Brown-backed, **251**.
Oyster-Catchers, The, **195**, **251**.
Ox-Bird, **217**.
Ox-eye (Black-breasted Plover), **237**.
    Meadow, **215**.
    Sand, **216**.

Pale-Breast, **239**.
Parrot, Sea, 7, **18**.
Partridge (Bob White), **261**.
    (Ruffed Grouse), **272**.
    Birch, **272**.
    Black, **270**.
    Night, **201**.
    Spruce, **260**, **270**.
    Swamp, **270**.
    Virginia, **261**.
Partridges, New World, **259**.
    Old World, **259**, **262**.
Peacock, **260**.
Peep (Least Sandpiper), **215**.
    (Semipalmated Sandpiper), **216**.
Peet-weet, **232**.
Penguin, **21**, **23**.
Petrel, Fork-tailed, **71**.
    Leach's, 66, **71**.
    White-rumped, **71**.
    Wilson's Stormy, 66, **68**.
Petrels, The, **65**, **66**.

Pewee (Woodcock), **201**.
Phalarope, Northern, **192**, **197**.
    Wilson's, **192**, **196**.
Phalaropes, The, **191**, **196**.
Pheasant, **261**, **272**.
    Water, **91**.
Pheasants, The, **260**.
Pigeon, Band-tailed, **293**, **296**.
    Passenger, **293**.
    Prairie (Bartramian Sandpiper), **229**.
    Prairie (Golden Plover), **239**.
    Sea, **20**.
    White-collared, **296**.
    Wild, **294**.
Pigeons, The, **293**, **294**, **330**.
Plover, Belted Piping, **194**, **247**.
    Black-bellied, **237**.
    Black-breasted, **194**, **237**.
    Chicken, **249**.
    Field (Bartramian Sandpiper), **229**.
    Field (Golden Plover), **239**.
    Golden, **194**, **239**.
    Grass, **229**.
    Green, **239**.
    Highland, **229**.
    Kildeer, **194**, **242**.
    Mountain, **247**.
    Piping, **194**, **245**.
    Red-legged, **249**.
    Ring, **244**.
    Ring-necked, **194**, **244**.
    Ruddy, **219**.
    Semipalmated, **194**, **242**.
    Swiss, **237**.
    Upland, **193**, **229**.
    Whistling Field, **237**.
    Wilson's, **194**, **247**.
Plovers, The, **191**, **194**.
Pochard, American, **114**.
Poke (American Bittern), **157**.
    (Little Green Heron), **152**, **164**.
Ptarmigan, **348**.
Puffin, 7, **18**.
    Masking, **18**.
Puffins, The, **5**, **6**, **18**.
Purre, **217**.

Quail, **260**, **261**.
    California, **266**.
    Florida, **266**.
    Old World, **259**, **262**.
    Texan, **266**.
Quaily, **229**.
Quawk, **152**, **166**.

Rail, Big, **177**.
    Blue, **184**.
    Carolina, **173**, **180**.
    Clapper, **173**, **177**.
    Common, **180**.
    King, **173**, **178**.

Index

Rail, Lesser Clapper, 179.
    Little Black, 173, 183.
    Little Red, 179.
    New York, 182.
    Red-breasted, 178.
    Sora, 173, 180.
    Virginia, 173, 179.
    Yellow, 173, 182.
    Yellow-breasted, 182.
Rails, The, 171, 172, 177.
Ring-neck, Pale, 245.
Robin, Beach, 209.

Sabre-bill, 234.
Saddle-back, 37.
Sanderling, 193, 219.
Sandpeep, 215.
Sandpiper, Ash-colored, 209.
    Baird's, 193, 214.
    Bartramian (Upland Plover), 193, 229.
    Black-bellied, 217.
    Bonaparte's, 213.
    Buff-breasted, 193, 231.
    Green, 225.
    Least, 193, 215.
    Long-legged, 208.
    Pectoral, 193, 212.
    Purple, 211.
    Red-backed, 193, 217.
    Red-breasted, 209.
    Robin, 209.
    Schinz's, 213.
    Semipalmated, 193, 216.
    Solitary, 193, 225.
    Spotted, 193, 232.
    Stilt, 193, 208.
    Swimming, 196.
    Western Semipalmated, 193, 217.
    White-rumped, 193, 213.
Sandpipers, The, 191, 192, 201.
Saw-bill (American Merganser), 87.
    (Red-breasted Merganser), 89.
Scissor-bill, 31, 59.
Scolder, 125.
Scooper, 198.
Scoter, American, 85, 130.
    Black, 130.
    Surf, 85, 131.
    White-winged, 85, 131.
Sea Swallows, The, 46.
Shag, 77.
Shearwater, Common Atlantic, 66.
    Greater, 66.
    Wandering, 66.
Shearwaters, The, 65, 68.
Shelldrake (American Merganser), 87.
    Buff-breasted, 87.
    Hooded, 91.
    Pied, 89.
    Red-breasted Merganser, 89.

Short Neck, 212.
Shoveler, 84, 107.
Shuffler, 118.
Sickle-bill, 234.
Skimmer, Black, 31, 59.
Skimmers, The, 31, 59.
Skua, Buffon's, 34.
Skuas, The, 29.
Snipe, 191, 192, 201.
    American, 204.
    Big-headed, 201.
    Blind, 201.
    Brant, 217.
    Checkered, 249.
    Common, 204.
    Cow, 212.
    Deutscher, 206.
    English, 204.
    Fall, 217.
    German, 206.
    Grass, 212.
    Gray (Dowitcher), 206.
    Gray (Knot), 209.
    Horsefoot, 249.
    Jack (Pectoral Sandpiper), 212.
    Jack (Wilson's Snipe), 193, 204.
    Little Stone, 224.
    Meadow, 212.
    Mud, 201.
    Prairie, 229.
    Quail, 206.
    Red-breasted, 206.
    Robin (Dowitcher), 206.
    Robin (Knot), 193, 209.
    Rock, 211.
    Sea, 196.
    Stone, 223.
    Surf, 193, 219.
    Telltale (Greater Yellowlegs), 223.
    Telltale (Yellowlegs), 224.
    Wall-eyed, 201.
    Whistling, 201.
    White (American Avocet), 198.
    White (Black-necked Stilt), 199.
    Wilson's, 193, 204.
    Winter, 209.
    Winter (Red-backed Sandpiper), 217.
    Wood, 201.
Sora, 173, 180.
Sorce, 180.
South Southerly, 85, 125.
Speckle-belly, 134.
Speckled Belly, 281.
Spike-tail, 281.
Spoonbill, 107.
    Roseate, 156.
Spoonbills, The, 151.
Squealer, 239.
Stake Driver, 157.
Stib, 217.

Index

Stilt, Black-necked, 192, 199.
Stilts, 192, 198.
Stint, 215.
    Wilson's, 215.
Stork, Wood, 155.
Storks, 151, 155.
Striker, Gannet, 47.
    Little, 54.
Striped-head, 235.
Swallow, Sea, 49.
Swallows, Sea, 31.
Swan, American, 143.
    Trumpeter, 86, 145.
    Whistling, 86, 143.
Swans, The, 86, 143.
Swimmers, Fully webbed, 75–79.
    Lamellirostral, 83–145.
    Long-winged, 29–61.
    Plate-billed, 83–145.
    Totipalmate, 75–79.
    Tube-nosed, 65–72.

Tattler, Bartram's, 229.
    (Greater Yellowlegs), 223.
    Semipalmated, 227.
    (Solitary Sandpiper), 193, 225.
    Wood, 225.
Teal (*see* Duck).
Teaser, 32.
Teeter, 232.
Teeter-tail, 232.
Telltale, 223.
    Lesser, 224.
Tern, Arctic, 31, 53, 55.
    Black, 31, 47, 56.
    Caspian, 48.
    Cayenne, 47.
    Common, 31, 49, 55.
    Gull-billed, 46.
    Least, 31, 54.
    Marsh, 31, 46.
    Paradise, 52.
    Roseate, 31, 52, 55.
    Royal, 31, 47.
    Short-tailed, 56.
    Silvery, 54.
    Sooty, 57.

Tern, Wilson's 31, 49.
Terns, The, 29, 30, 35, 46, 48, 50.
Tildillo, 199.
Tilt, 199.
Tilt-up, 232.
Timber Doodle, 201.
Tinker, 23.
Toad-head, 239.
Turkey, Colorado, 155.
    Mexican, 260, 289.
    Rio Grande, 289.
    Water, 155.
    Wild, 260, 288.
Turkeys, The, 260, 288, 289.
Turnstone, 194, 249.
Turnstones, 194, 249.

Vulture, Black, 301, 307.
    Turkey, 301, 304, 307.
Vultures, American, 301, 304.

Wavey, 136.
Weaser, 87.
Whimbrel, 235.
Whistle Wing, 121.
Whistler, Brass-eyed, 121.
    (Golden-eye Duck), 85, 121.
    (Red-breasted Merganser), 89.
White Back, 116.
White Belly, 281.
Widgeon, American, 84, 100.
    European, 102.
    Wood, 111.
Willet, 193, 227.
Witch, Water (Horned Grebe), 9.
    Water (Pied-billed Grebe), 11.
Woodcock, 193, 201.

Yellowlegs, 193, 224.
    Big, 223.
    Greater, 193, 223.
    Summer, 224.
    Winter, 223.
Yellow Shanks, 223.
Yellow Shins, Lesser, 224.
Yelper (Yellowlegs), 224.
    (Greater Yellowlegs), 223.

THE POPULAR ORNITHOLOGY   BY NELTJE BLANCHAN

A boxed edition of "Bird Neighbors" and "Birds that Hunt and are Hunted." These two volumes cover all of our well-known birds. Text by Neltje Blanchan, annotated and with introductions by John Burroughs and G. O. Shields. The 100 colored plates present an unexampled series of bird pictures, being colored photographs from the birds themselves. 2 vols., octavo, boxed, $4.00.

JUST ISSUED

Birds that Hunt and are Hunted

Annotated and with an introduction by G. O. Shields, "Coquina"; a companion volume to "Bird Neighbors." This new book gives the life histories more completely than ever before in a popular work of 170 of our game and water birds and birds of prey, and contains 48 colored plates.

SPECIFICATIONS :—*Size, 7¾ x 10¾ ; Pages, 250 ; Strong green cloth binding ; Type, 11-point; Printed on fine paper, full margins for notes. 48 colored plates. Price, $2.00.*

Bird Neighbors

WITH INTRODUCTION BY JOHN BURROUGHS

"*Through certain unique features this book makes the identification of birds simple and positive even for the inexperienced.*" — *Boston Times.*

20th Thousand

"*The fifty colored plates in this truly sumptuous volume are most beautiful and accurate. . . . Such books as this one add new interest to life.*"—*Scientific American.*

An introductory acquaintance with 150 of our common birds, with 52 superb full-page pictures in color photography from the birds themselves. John Burroughs, the highest authority on this subject, who has annotated the text, says of the colored plates, in his introduction:

"*When I began the study of birds I had access to a copy of Audubon, which greatly stimulated my interest in the pursuit, but I did not have the opera glass, and I could not take Audubon with me on my walks, as the reader may this volume, and he will find these colored plates quite as helpful as those of Audubon or Wilson.*"

This book makes the identification of our birds simple and positive, even to the uninitiated, through certain unique features.

SPECIFICATIONS :—*Size, 7¾ x 10¾ ; Pages, 250 ; Strong green cloth binding ; Type, 11-point; Printed on fine paper, full margins for notes. 52 colored plates. Price, $2.00.*

DOUBLEDAY & McCLURE CO., Publishers,
141-155 East 25th Street, New York.

TWO OTHER CHARMING NATURE BOOKS

The Butterfly Book
By W. J. Holland, Ph.D., D.D., LL.D.

This is "A Popular Guide to a Knowledge of the Butterflies of North America," telling of their life and habits, and how they may be identified, collected and studied. Dr. Holland is Chancellor of the Western University of Pennsylvania, Director of the Carnegie Museum of Pittsburg, Fellow of the Zoölogical and Entomological Societies of London, Member of the Entomological Society of France, etc., etc.; he is, moreover, perhaps the first authority in America on this subject, and he has certainly the largest and most perfect collection in existence of the butterflies of North America.

The superb colored plates, photographed directly from the author's specimens, represent the highest mark yet made in color photography, and with the many text illustrations figure practically every species in the country; and the book is the first successful attempt to popularize this fascinating pursuit.

SPECIFICATIONS:—*Size*, 7⅝x10½; *Pages*, 350; *Binding*, cloth, with butterfly stamped in color; *Type*, 11-point French Elzevir; *Illustrations*, 48 plates in color and many text illustrations, showing every important American species. *Price*, $3.00, net.

* * *

Flashlights on Nature
By Grant Allen

A book describing the life histories of our most familiar insects and plants, and calling attention in the author's charming style to many neglected miracles in our everyday world. The illustrations include more than 100 drawings by Frederick Enock, who has made his sketches directly from life under the microscope; he has often watched, as Mr. Allen tells us, for twelve hours at a time to intercept a rare chrysalis at its moment of bursting, or for the favorable moment for drawing some unusual characteristic.

SPECIFICATIONS:—*Size*, 5⅜x8; *Pages*, about 300; *Fully Illustrated*; *Binding*, cloth, decorated; *Type*, 11-point. *Price*, $1.50.

DOUBLEDAY & McCLURE CO., Publishers,
141-155 East 25th Street, New York.

www.ingramcontent.com/pod-product-compliance
Lightning Source LLC
Chambersburg PA
CBHW022109300426
44117CB00007B/649